STATISTICAL GEOGRAPHY

STATISTICAL GEOGRAPHY

METHODS AND APPLICATIONS

ZAMIR ALVI

RAWAT PUBLICATIONS

Jaipur • New Delhi • Bangalore • Guwahati • Kolkata

ISBN 978-81-7033-270-1 (HB)

Reprinted, 2023

Published by
Prem Rawat for **Rawat Publications**
Satyam Apts., Sector 3, Jawahar Nagar, Jaipur - 302 004 (India)
Phones: 0141 265 1748 / 265 7006
E-mail : info@rawatbooks.com
Website: rawatbooks.com

New Delhi Office
4858/24, Ansari Road, Daryaganj, New Delhi 110 002
Phone: 011-23263290

Also at *Bangalore,Guwahati* and *Kolkata*

Typeset by Rawat Computers, Jaipur
Printed at Nice Printing Press, New Delhi

FOREWORD

One of the major developments in geographical studies in the last two decades is the application of statistical methods in order to achieve precise results for planning purposes. These methods help in the proper interpretation and understanding of complex phenomena. If geography is to explain the human behaviour or reaction of physical, social and economic factors, it will not suffice to describe these differences, but one has to resort to principles or if possible laws, to explain them and then one has to lean on the various statistical methods for a cogent analysis and interpretation for a clear understanding.

There is however, a general reluctance particularly among arts and social science students, towards the application of the various statistical methods owing to their poor knowledge in mathematics or statistics. David Harvey, the author of 'Explanations in Geography' states that initially, not being well-versed in statistical methods, often found that he could not interpret the results of his own analysis because of his lack of command of statistics and mathematics, but in due course of time he did realise the value of statistical methods when he says 'we can formulate criteria to judge whether or not an argument is sound, a technique properly used or an explanation reasonable'.

In the analysis of phenomena, one must know as unambiguously and precisely as possible what one is investigating. The topic or the particular aspect of that topic being investigated is termed the *dependent variable*. It is called a variable because it varies in quality or quantity and it is dependent variable because we are interested in knowing how its quality or quantity varies with changes in the quality or quantity of other phenomena which we think are related to it. The other phenomena are called *independent* variable. A particular phenomenon may be a dependent variable in one case and an independent variable in others. When we examine the spatial variations of the dependent variable over the study area, we can come to meaningful conclusion in respect of planning.

There is no dearth of books on statistical geography, but the merit of the present volume is that the author of the book, Professor Zamir Alvi, has been teaching the subject to post-graduate classes for a long time and realizing the difficulties of geography students in the application of statistical methods, he has presented the methods in this volume in a simple and digestable manner and hedging them with examples and their solutions, made them easy and comprehensible.

I am sure the book will be widely received by the students of geography and all those interested in the application of statistical methods.

Aligarh

Mohd. Shafi
Professor Emeritus
Department of Geography
Aligarh Muslim University
ALIGARH

PREFACE

The use of statistical techniques in geography is rather a recent development. In present day geography, therefore, great weightage is being given to statistical applications. To encourage and make proper use of statistical techniques in geography, a number of books, some of them very good ones, have been published. But in most of these books the authors have assumed that the student knows the derivations and some steps have therefore been skipped over. Generally it has been observed that the students of geography do not have a good background of mathematics. The steps therefore in the derivations, having been skipped over, make it difficult for the students to follow the particular technique.

Substituting the values in a formula and getting some result, does not serve the purpose. One must understand the implications of the result thus obtained. In view of the lack of basic mathematical background among geography students, the author in this book has tried to proceed step by step in order to make various statistical problems more easily understandable.

After the introduction, the first chapter deals with Measurements in Geography followed by Frequency Distribution, Measures of Central Tendency, Measures of Dispersion, Probability Distribution, Sampling, Hypothesis Testing, Correlation and Regression and Elementary Introduction to Geographic Matrices. In all these chapters as far as possible, illustrations have been given which help in the better understanding of the techniques.

Zamir Alvi

ACKNOWLEDGEMENTS

I would like to express my sense of gratitude to Professor S.M. Rafiullah, who inspired me to write a book on the use of statistical techniques in geography and was always available for help when the need arose. I am very much thankful to Dr. Jamil Ahsan and Dr. Mohammad Yaqub of the Department of Statistics, Aligarh Muslim University, who went through parts of the script and gave me valuable suggestions. Without their assistance this book could not have been published. I am specially thankful to Professor Mehdi Raza the then Chairman of the Department for making available to me the relevant material from the Seminar Library of the Department at the time when I started writing this book. I am also thankful to Professor Mohammad Shafi and Professor Abdul Aziz, the former Chairman of the Department, who gave me valuable suggestions. I am greatly indebted to my students in most of whom seeing the weak background of basic mathematical understanding, I found it necessary to help them by bringing out this book.

Zamir Alvi

CONTENTS

INTRODUCTION

The use of statistical methods in geography is the most recent development. Statistical Geography has been considered as a new branch of geography. Statistics in itself is a recent development which is related to mathematics—a very old discipline. Although closely related to mathematics, statistics is a very important field by itself. Mathematics is a science which is the most fundamental of all sciences. It deals with the behaviour and function of numbers. It deals with certainities of which it gives proof e.g. $(a+b)^2=a^2+b^2+2ab$ can be proved mathematically with the help of Fig. 1.

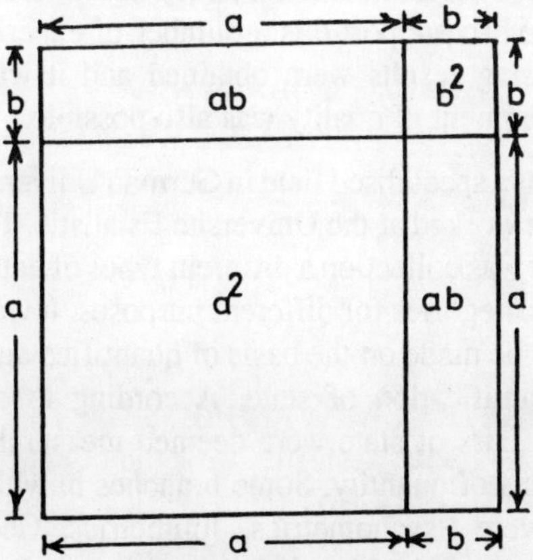

Fig. 1. Geometrical proof showing
$(a + b)^2 = a^2 + b^2 + 2ab$

On the other hand, statistics deals with uncertainities and possibilities e.g. if a coin is tossed, sometimes head and sometimes tail will appear. One cannot say with surity as to which face would appear. There is a matter of chance.

Statistics actually started with gamblers. Bids were made and money was put on stakes. Various types of methods of tips were tried. But in academic field it was first of all introduced in seventeenth century by some British political mathematicians like Graunt and Petit. It was pointed out by Meitzen in the middle of nineteenth century that even the social affairs can be expressed in terms of quantitative measurement. Gally and Petit also held the same opinion. In the beginning of eighteenth century, in a German university a scholar named Achenwal said that the category of states may be expressed and determined in terms of quantitative measurement. Gally and Petit also held the same conviction. In the beginning of eighteenth century, in German University a scholar named Achenwel said that the category of states may be expressed and determined in terms of quantitative measures. Another German scholar Leiserfeld also held the same opinion. The authority of the above statement is debatable because some statisticians think that the idea was put forward by Courney who was contemporary of Petit and Graunt and not by Achenwel.

There are two basic approaches of assessment—Qualitative & Quantitative. Qualitative assessment is very easy which can be made in quantitative form also. To support this a number of experiments were performed and rousing results were obtained and it was supposed that quantitative assessment of quality was also possible.

Statistics was a specialised field in German University. A number of German scholars worked at the Universite Estatistie. The main purpose of the University was collection a different types of data their classification in different categories for different purposes. It was found out that a set of state can be made on the basis of quantified analysis. The main interest was quantification of state. According to certain laws and principles, the affairs of state were deemed measurable and could be expressed in terms of quantity. Some branches in which statistics was much utilized were Psychometrics, Julimetrics, Geometrics etc. In seventeenth century, British statisticians also introduced it and hence there was a competetion with the Germans resulting in very acute criticism because quantification intervened in cultural fields also.

The modern trend in every discipline is that the method should be made quantity based. Statistics means measurement of certain characteristics of the item of study. It involves not only vision but drawing of such a state also. Statistical Geography means the use of methods of statistics in the study of Geography. It was in the middle of twentieth century that Geography also started its application in methods developed in statistics during the last two centuries. Therefore to keep ourselves abreast with the revolution in the world of Geography, one must have acquaintence with it. It is therefore nowadays being used in different branches of Geography like Geomorphology, Human Geography, Economic Geography, Agricultural Geography etc.

The latest development is the increasing application of statistical methods and the refinement of measures. Therefore theoritical and basic methods which would apply to Geography should be carefully understood.

Quantification revolution began after the Second World War but they have become noticeable only recently and therefore the whole of Geography is being reconstructed on the basis of quantification which means the use of quantity.Simple forms of quantification has been going on for a very long time e.g. in counting measurement, classification, categorization, numerical analysis, development of hypothesis, theories, principles etc. Collection of data has always been there but categorization, measurement etc. is a new thing. Quantification has been introduced in Geography by non-geographers e.g. Botanists, Ecologists etc. They have shown that a number of geographical problems could be solved by this new approach and hence even Geographers are now paying attention towards quantification.

The main features of qualification are:

1. Collection of data
2. Processing of data
3. Plotting of data and making a model so that certain patterns come out
4. Comparison of these models and assessment of actual situation
5. Repeating comparisons under diversified conditions
6. Generalization and making hypothesis
7. Formation of theory and
8. Formation of law.

Example

Suppose some industry is to be set up.

1. For the industry to be set up the following data should be collected:

   (a) Raw materials required

   (b) Quantity of raw materials to be used

   (c) Number of persons (skilled and unskilled) required

   (d) Source of supply of raw materials

   (e) Means of transport and

(f) Market (at home and abroad) etc.

2. and 3. For establishing the industry, the ideal conditions must be present for starting it. Some of these conditions have been indicated above just by factual statement. If the data conforms with that in a well established industry, it means that the industry is ideally located. If there is too much unconformity, reason should be known from the model thus prepared.

4. The actual industry should be compared with the model.

5. Apply that model to a number of cases spread over bigger region.

6. After that experimentation and generalization of conclusion so that a hypothesis is formed.

7. When the hypothesis develops, it is tested and repeatedly tested in different periods of time under diversified conditions and if it suits, it becomes a theory .

8. The theory is still tested in diversified conditions all over the world in various periods of time. If it becomes successful, it becomes a law.

Geography is now remodelled on the above basis. Hypothesis may be discarded if it does not hold good but if it holds good it becomes a theory. A theory may also be discarded or it may become a law. Every branch of science holds a law and if Geography is to be considered as a subject of science, it must also hold a law.

CHAPTER I

MEASUREMENTS IN GEOGRAPHY

The basic root of numbers is unity. Quantities are of two types — Discrete and Continuous. Four chairs can be concieved but $4\frac{1}{2}$ or $7\frac{3}{4}$ chairs cannot be concieved. The chairs are, therefore, described as discrete quantity. In other words a number without a fraction is a discrete number. But if the fraction is meaningful, the quantity is a continuous quantity i.e. where fraction can be concieved e.g. distance, age etc. Discrete quantity is expressed in integers whereas a continuous quantity is expressed in integers and fraction both. Continuous quantity cannot be used for discrete quantity. Therefore the first thing to be noted is that the data requires continuous or discrete numbers. Both the numbers are not mutually exclusive. They are at times related to one another e.g. the speed of a train can be 60 kms. per hour as well as 60.5 kms. per hour. It means that the speed can be represented in continuous as well as in discrete form. When an item is expressed in continuous form, it can be expressed in discrete form also but the reverse is not correct.

The length of a fleet of ship can be expressed in discrete form only because the ships have to be in whole numbers and not in halves and fractions. Discrete numbers cannot be expressed in discrete form only when they are not advised to be shown thus.

NUMERICAL SYMBOLS

There are two sets of numerical symbols Arithmetic and Algebraic. 1, 2, 3, ... are arithmetic symbols whereas a, b, c, ... are algebraic symbols.

One (1) is a unit which indicates a single individual. It is not specified whether it is for one horse, one man or what? Two (2) means one individual added to one individual. Similarly three (3) means 1 + 1 + 1. 1, 2 and 3 are all arithmetic symbols. They have a rigid meaning i.e. 1 çannot be more than 1, or 5 and 6 cannot be more than 5 and 6 respectively.

a, b, c ... or x, y, z etc. are Algebraic symbols. They may stand for a particular quantity assuming any arithmetic meaning. 4 + 5 = 9 but 4 chairs and 5 tables cannot be equal to 9 chairs, tables because chairs and tables cannot be added together. If 'a' stands for chairs and 'b' stands for tables, then the total is a + b, i.e. a set of 'a' and 'b'. With a, b, c or x, y, z, only generalization is possible which can assume any arithmetic equivalent.

Suppose there are 10 persons with their ages ranging from 1 to 5 years. This population can be broken up in age groups.

Let P stand for population

Let P_1 stand for population of persons of 1 year age;

Let P_2 stand for population of persons of 2 year age;

Let P_3 stand for population of persons of 3 year age and so on

i.e. the sum total becomes $P_1 + P_2 + P_3 + P_4 + P_5$.

If three children are one year old, it means $P_1 = 3$ and so on.

In other words this type of structure is possible for any population i.e. $P_1 + P_2 + P_3 + P_4 + \ldots\ldots + P_n$.

In case of Arithmetic symbols, this type of structure is not possible.

The most exclusively used symbols are (+) plus; (–) minus; (×) multiplication and (÷) division. Other less commonly used symbols are raising and extraction of power e.g. 3^2 means 3 raised to the power 2 and √ means square root and $\sqrt[n]{P}$ means nth root of P which can also be written as $(P)^{1/n}$. This form has an extra advantage specially for logarithmic calculations. One of the most widely used symbol in Statistics is a Greek letter sigma (Σ) which stands for summation i.e. instead of writing $P_1 + P_2 + P_3 + \ldots P_n$, we can write $\sum_{i=1}^{n} P_i$ i.e. summation of P series from 1 to n when the mean difference is 1.

NATURE OF GEOGRAPHIC DATA AND PROBLEMS OF KINDS OF MEASUREMENTS

The main purpose of introducing Statistics in Geography is to quantify and express geographical elements in terms of numbers. The geographical elements may be the distribution of population, vegetation or densities.

Data relates to areas or points on the surface of the earth or relationships or aspects of relationship between two sets which are very much used in geographical studies. The first aspect of nature relates to areas, lines or points. It is not purely a characteristic of geography alone because such measurements are used in other disciplines also specially geometry i.e. areas, points etc. In geography relevant methods are borrowed specially from geometry which is not difficult. But the difficulty arises when we take up the job of measuring or quantifying attribute and relationship between elements of factors of study. The critics of quantification have taken a big step at this situation and there develops the anti-quantification thesis primarily on the basis that since geography is considered to be a science of relationship and relationships and attributes occuring from relationship cannot be directly measured or quantified therefore quantification in geography is not possible and any attempt at quantification would be not more than an academic exercise. It is partially but not totally valid i.e. measurement of characteristics or attributes or quality cannot be so directly measured as the size or the value or the frequency of the elements that are observed or observable on the surface of the earth. But it is practically invalid also because the recent and more modern developments in mathematics and statistics have made it clear that even such matters as attributes and qualities may be rendered into measurable forms in a special way suitable to the problem. Thus one can be contended with the fact that quantification of the aspect of the nature of relationship can also be done but of course not in the same traditional style as in the measurement of the size, value or the frequency. On the whole it can be argued that the quantification criticism now stands rather on weak ground and that simply because of a different style of measurement required to be applied to the qualitative aspect of relationship, one cannot and should not rule out quantification in geography and hold the view that quantification is not possible.

'High', 'Low' or 'Medium' are apparently quantified representation but basically qualitative because their quantitative equivalents are not known. Suppose the ranges are as follows:

5%	to 10%	Very Low
10%	to 15%	Low
15%	to 20%	Medium
20%	to 25%	High

The above table with the ranges given, makes the representation distinctly clear. But the same item studied in any other part of the world may show the very low percentage as 40 to 50; medium as 50 to 60 and high as 60 to 70. Very low in this case is higher than high in the previous case. The techniques therefore should be specific and more well-defined which is a part of statistics. 'High', 'Low' and 'Medium' seem to have the same representation but since their values are not known, one cannot be sure as to what do they stand for ? For similar other problems, statistics is used to resolve them in quantitative form.

The data is of various nature which has to be clearly understood with respect to its properties, requirements etc. Geography is an area bound study and geographic data essentially an areal data. Two broad aspects of areal data have been recognized:

1. Data relating to space or a segment of space;
2. Data relating to the nature of characteristic of a single item or a number of items studied and found in that particular segment.

The table on the right hand side shows areal data as a space segment (Aspect 1). The amount of rainfall is a numerical value which is another set and an example of Aspect 2. As aspects 1 and 2 are combined, a certain representation is obtained.

Rainfall

Low	High
Medium	Very high
Low	High

In space component data, it is difficult to define and firmly specify a unit of that data. Unit is the base which is fundamental for the development of any item. The state of Uttar Pradesh may be taken as a unit. It is a big state. Uttar Pradesh may be divided in 'Tehsils',

'Parganas' and villages depending upon the precision required. Because of space being continuous, it is not defined in units. 1, 2 and 3 are firmly defined. The world as a whole or any of its region is continuous which is a unit but it cannot be firmly defined.

The space segment may be expressed in three ways

1. By expressing area in terms of areal measurement.

2. By expressing in terms of administrative unit i.e. in terms of a continent or a political or administrative unit or India, Uttar Pradesh i.e. sub-unit or sub-sub-unit.

3. By expressing in terms of other well-defined and adopted attributes i.e. a geographical region e.g. the amount of rainfall in a particular region is suppose 40 to 50 millimetres but the area in which that amount of rainfall has been experienced is not indicated.

Area as a space segment does not have any single item. It may have soil, rainfall, landform, vegetation, minerals etc. Therefore one or more than one of these items are selected to make a study. If a single item is taken, the problem becomes easy. If more than one item is taken, it is a little bit complicated. Three or more than three items cannot be easily combined unless one is very well acquainted with well defined techniques of combining them for which advanced knowledge of statistics is essential.

The items may have varied types of data e.g.

1. Value Type
   - (a) Absolute value e.g. 500,000 persons.
   - (b) Ratio value e.g. 50% of world output.
   - (c) Ranked or ordered value e.g. 3rd in the world.

2. Attribute Type
   - (a) Discontinuous attribute e.g. sex ratio, rural-urban, male-female ratio in which no third item is possible.
   - (b) Multiple natured attribute e.g. Linguistic groups, age

composition, occupational groups-all of which have more than two components.

(c) Static Attribute where aspects donot change i.e. phenomenon of births, deaths, fertility, mortality etc. The numbers may change but the phenomenon is constant or e.g. the area of the world or Physical Divisions which are more or less a constant phenomenon in one's lifetime.

(d) Dynamic Attribute where aspects change every moment e.g. population, age, age-composition are all continuous changes which cannot be solved

When the values and attributes are combined, every sub division of one set can be combined with every sub division of the other set.

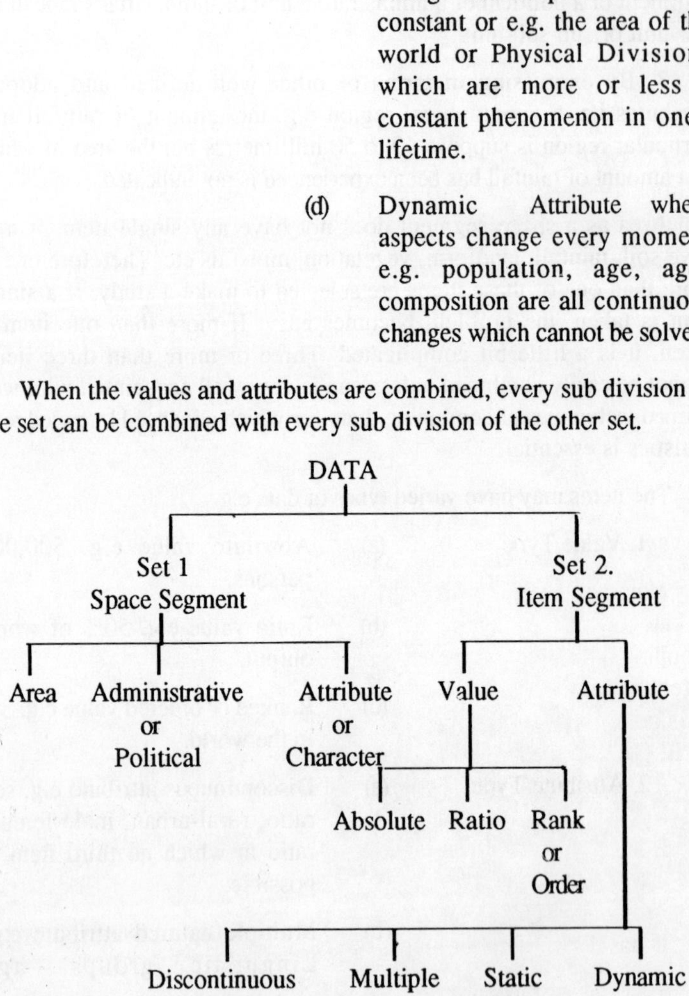

The area in Set 1 can be combined with value in Set 2 or even in rank or order of the sub division in Set 2. Unless unbiased techniques are adopted, any definite conclusion cannot be reached.

Having defined the nature of data, one must know as to how best the data can be expressed. Length is a space which is the shortest distance between two points. What should be the unit of length — yards, feet, metres etc ? There are varieties of measures but length is measured with respect to the nature of work going to be taken i.e. the best measure which suits the nature of our data or the structure of our data.

Two basic components for effectively measuring geographic data are—area and aspect. Measurements of aspects generally do not differ much in other sciences. Typical aspects of Geography are not found in other disciplines, so how to measure that? Various techniques have to be applied to fulfill the purpose of investigation in a way such that it is suitable for an area or location component of investigation.

The four main scales of measurements depending upon the type of investigations desired are as follows:

a) Nominal Measurements

b) Ordinal Measurements

c) Interval Measurements and

d) Ratio Measurements.

a) Nominal Measurements

Nominal measurement is the least quantitative in character. It refers to categorization i.e. it is a classificatory measurement referring to grouping of data in categories or classes. It can be quantified upto some extent but on it arithmetic operations are not easy to be performed satisfactorilly e.g. classification of towns on the basis of its size e.g. A, B, C and D sized towns.

A has 5 divisions forming a town

B has 4 divisions forming a town

C has 6 divisions forming a town and

D has 3 divisions forming a town.

The simplest form of measurement is zero-one system. In Fig. 2,

A is connected to C by road

A is connected to D by road

B is connected to C by road

B is connected to D by road and

A is not connected to B by road.

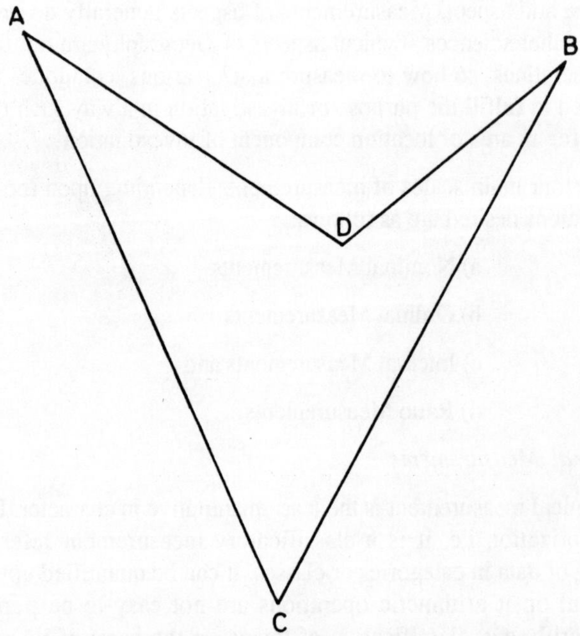

Fig. 2. Connectivity and Non-connectivity in the town A, B, C and D

The towns if connected are grouped as 1 category e.g. AC – 1, AD – 1, BD – 1, BC – 1.

The towns if not connected are grouped as zero category e.g. AB – 0.

One and Zero categories show only the presence of a town. It does not have a numerical value. It only shows whether they are connected or not connected. Zero means an absence of connection (not connected) whereas 1 shows connection. It is indicated symbolically as | 0, 1 | which is an example of connectivity measure. A town is considered as a point having a fixed location. It is not a point in geometric sense. The main purpose of point data is to study connectivity i.e. connection or non-connection between two towns or similar other cases. In the following figure, A is not connected with A and B is not connected with B i.e its category is zero. It may be illustrated thus:

	A	B	C	D
A	0	0	1	1
B	0	0	1	1
C	1	1	0	0
D	1	1	0	0

ZERO-ONE SYSTEM OF CONNECTIVITY OF TOWNS IN NOMINAL MEASUREMENT

The above system may be applied in other similar situations e.g. problems of students in attendence—presence or absence. The table may be plotted as shown above. It leads to geographical matrix which is an arrangement of data in rows and columns. Matrix is just like a box of which each mesh has been assigned a certain data i.e. an array of data called matrix.

b) Ordinal Measurements

Ordinal measurement refers to ranking or establishing order of the elements of items of study. The order may be ascending or descending. The value of order i.e. 3rd or 5th is only symbolic and is not receptive of arithmetical operations. Suppose A, B, C and D are four towns and the rank or order is as follows:

A	B	C	D
2	3	1	4

In it 1 may be considered as universe and 2 a part of it.

Towns	C	A	B	D
Ranks	1	2	3	4

Ranks are not receptive of arithmetic operation e.g. to say that C + A = 1 + 2 = 3 or C + A = B is not correct. A, B and C are only symbolic.

1. Rank or order should always be asymmetric e.g. if C is greater than A, then A cannot be greater than C. Asymmetric means at the same time i.e. any item cannot be greater than and smaller than the other item simultaneously.

2. Rank or order is transitive i.e. rank values are from one item to another item in ascending or descending order e.g. if A is greater than B and B is greater than C it means that A is greater than C i.e. greatness is transiting in descending order.

Illustration:

Tehsil	Number of Farms	Order
1	35	1st is 35
2	28	3rd is 30
3	30	2nd is 28
4	15	4th is 15

The order column is in descending order which indicates higher position in ordinal measurement. The above table can be arranged as follows:

Number of Farms	Ranks
35	1
30	2
28	3
15	4

The table now indicates the status of tehsils one after the other. For ordinal scale, arithmetic operation should never be applied.

c) Interval Measurements

Interval measurement is a measurement in which the space which separates two numbers or items of study is specified. The state of separation may be of kind or space. In it zero is not defined but it is arbitrary e.g. in Centigrade or Fahrenheit thermometres, zero is not well defined.

Interval may be defined in accordance with the nature of investigation e.g. continuous interval scale or discontinuous scale. A 50 year rainfall data may be divided in interval which is not there in the data. Interval data may be ordinal as well as nominal measurement in which zero point has to be defined. Arithmetic operation can be utilized on a particular scale of interval e.g.

on 8th April, the temperature was 70° F } I
on 9th April, the temperature was 68° F }

The difference in temperature is 2° F

Suppose on 8th April, the temperature was 70° F } II
and on 9th April, the temperature was 50° C }

The difference in Case II cannot be found by subtracting one from the other because zero is not defined i.e. the value of zero in both the cases is different.

d) Ratio Measurements

Ratio measurement is most fundamental because zero is well-defined. It is the true value and refers to proportions or ratios between two items different or alike. On this system all arithmetic operations can be applied and on which comparability is maximum. Ratio refers to use of arithmetic number in full meaning.

TABULATION OF DATA

A Table in one span provides too many informations which can be arranged and interpreted in a number of ways in accordance with the requirements. A Table is defined as the simplest statistical technique not only in geography but in many other fields also. A Table is also defined as a list of individual values or of the sets. In a Table no order of recording is prescribed and no arrangement is required. Such a simple form of Table is the matter used as a raw material for further interpretation, analysis and for preparing more complex Tables. Such a table is called a 'Summary Table'. Summary tables are more defined and more revealing instrument of assessment and analysis.

The parts of a table are first, number of a table followed by heading and sub-heading e.g. in Table I, the district is divided in 'Workers' and 'Non-Workers' and then in males and females. After that are column heads. Then each head may be further sub-divided. In Table I, there are three spanners—Total, Workers and Non-Workers. The first column showing the names of the districts is called the 'Stud' of the table which shows the field of the table. The several parts of the table therefore may be its number, heading, sub-heading, columnar heading, spanners, sub-heads, studs and field.

The same table may be arranged in a number of ways in accordance with the requirements. The second column of the table is not arranged in descending or ascending order. If it were so, the table would be termed as Ordered table. If the table is prepared showing the names of the districts in alphabetical order, then it would be much easier to locate the name of the district.

Besides simple table the other type of table is Summary or Complex table which is also prepared in accordance with the work required to be handled. Some of the summary tables are very very complex. In Table I, there are six districts which may be called a region. The work force of the region as a whole can be found out by adding the six figures in the fourth column. The observer may focus his eye on the totals showing dependent or independent figures and reach a certain conclusion. If one is interested in female workers, their proportion to the total workers could be found out. The column may be further broken up into dependent and independent, rural and urban etc. which would make a still complex table. If percentages are calculated, their total should also be given which should always add up to 100.

The main object of a table is to present a numerical distribution and frequency of distribution of the elements of the universe. Frequency means number of occurrences whereas distribution means arrangement in space. Temporal frequency shows occurrence of elements in time. The time may be of varied types and hence the table can be made further complex.

CHAPTER II

FREQUENCY DISTRIBUTION

Frequency means number of occurences of individual values in a set of data. There are two principal forms of expressing frequency distribution:

a) Tabular Form and

b) Graphic Form.

a) Tabular Form

Tabular form of frequency distribution means arrangement of statistics in the form of a table either ordered or unordered.

TABLE II

Annual Rainfall at A

Year	Rainfall (mms.)
1940	290
1941	310
1942	280
1943	290
1944	300
1945	320
1946	310
1947	340
1948	280
1949	290
1950	300

TABLE 1

Labour Force

Districts	Workers			Non-Workers			Total		
	Male	Female	Total	Male	Female	Total	Male	Female	Total
Muzaffarnagar	504,040	13,626	517,666	480,455	804,168	1284,623	984,495	817,794	1802,289
Meerut	904,275	26,396	930,671	934,277	1502,005	2436,282	1838,552	1528,401	3366,953
Bulandshahr	537,561	14,195	551,756	580,880	940,707	1521,587	1118,441	954,902	2073,343
Aligarh	563,977	12,975	576,952	587,486	947,391	1534,877	1151,463	960,366	2111,829
Mathura	342,914	11,782	354,696	365,272	570,339	935,611	708,186	582,121	1290,307
Agra	612,299	14,288	626,587	649,956	1032,095	1682,051	1262,255	1046,383	2308,638

In Table II no order has been maintained as far as the magnitude of the data is concerned. The only order that is maintained is that the years are successive i.e. the stud is in chronological order but this table would not be called an ordered table. An ordered table in the true sense is that table in which the data is arranged on the basis of its magnitude.

In Table II some figures occur again and again e.g. the figure of 290 occurs thrice, 280 twice and so on. On that basis the same table can be summarised as shown in Table III.

TABLE III

Frequency Distribution
(Frequencies in ordered from)

Rainfall (mms.)	Frequency	Cumulative Frequency
290	3	3
310	2	5
280	2	7
300	2	9
320	1	10
340	1	11

Table III shows another form of frequency distribution in which the frequency is ordered.

The amount of rainfall from the same table can be arranged in graded form as shown in Table IV.

TABLE IV

Frequency Distribution
(Items in graded form)

Grades (mms.)	Frequency
280 – 290	5
300 – 310	4
320 – 330	1
340 – 350	1
Total	11

The frequencies in the preceding table should always add up to 11. This is another form of frequency distribution where the frequencies are ordered and the items are graded.

TABLE V

Annual Rainfall
(1911 to 1970)

Grades (mms.)	Frequency (Years)
250 – 300	15
300 – 350	11
350 – 400	12
400 – 450	10
450 – 500	6
500 – 550	6
Total	60

Table V can be modified in another form called 'More than' form as shown in Table VI.

TABLE VI

Frequency Distribution (More than Form)

More than (mms.)	Frequency
250	60
300	45
350	34
400	22
450	12
500	6
550	0

There exists a particular relationship between Table V and Table VI e.g. if the first figure of frequency in Table V is subtracted from the first figure of frequency in Table VI, the result is the second figure of frequency in Table VI and so on. The frequencies in Table VI cannot be easily shown by Table V.

With the help of ordinary table, the frequencies may also be arranged in 'Less than' form as shown in Table VII.

TABLE VII

Frequency Distribution
(Less Than Form)

Less Than (mms.)	Frequency
250	0
300	15
350	26
400	38
450	48
500	54
550	60

Tables VI and VII are also related e.g. if the frequency figures for 'Less Than' of 350 mms. are added to the frequency figures for 'More Than' of 350 mms., the result is 60 i.e. the total number of years for which the data has been recorded.

The preceding description indicates various ways in which the tables may be prepared and arranged in accordance with the requirements.

The data in grouped or graded form may be an Open-end or Close-end data. The Open-end data and the Close-end data have been shown in Tables VIII and IX respectively.

TABLE VIII

Annual Rainfall
1911 to 1970
(Open-end Data)

Grades (mms.)	Frequency (Years)
Less than 300	15
300 – 350	11

350 – 400	12
400 – 450	10
450 – 500	6
500 and Over	6

TABLE IX

Annual Rainfall
1911 to 1970
(Close-end Data)

Grades (mms.)	Frequency (Years)
250 – 300	15
300 – 350	11
350 – 400	12
400 – 450	10
450 – 500	6
500 – 550	6

In case of open-end data system of frequency distribution as shown in Table VIII, the first group can include any figure below 300 mms. e.g. 240 or 148 or only 7 mms. and similarly the last group can include any figure above 500 mms. e.g. 962 or 1764 or 32456 mms. etc. But in case of close-end data system, as is very clear from Table IX, the upper as well as the lower limits are very rigid e.g. nothing less than 250 mms. or more than 500 mms. can be included in this table.

In yet another form of frequency distribution the lower and the higher figures of each of the groups are over-lapping as shown in Table IX e.g. the figure of 300 mms. occurs in the first as well as in the second group and 350 mms. in the second as well as the third group.

When the figures are over-lapping, the first group in the above case will include figures from 250 mms. upto less than 300 mms. and 300 mms. will be included in the second group and so on. But in the case of non-over-lapping grouping system, similar situation does not arise.

(b) Graphic Form

The position of a point in a plane is defined by a certain type of system called Coordinate System. The position of any point can be defined in relation to any other object. In Figure 3, XX´ is a horizontal line called abscissa and YY´ the vertical line called ordinate. Their point of intersection O is called origin. If the position of any point P is to be found out, the distance along XX´ and YY´ will have to be measured. In Figure 3, the distance along XX' is 5 units and that along YY´ axis is 3. Therefore 5 and 3 are called the coordinates of the point P. They are written as (5,3). This system of location of a point is known as 'Rectangular Coordinate System'.

The graphic system of showing frequency distribution gives a visual expression of the object shown in the table. If properly interpreted, it also gives the trend which the graph will follow.

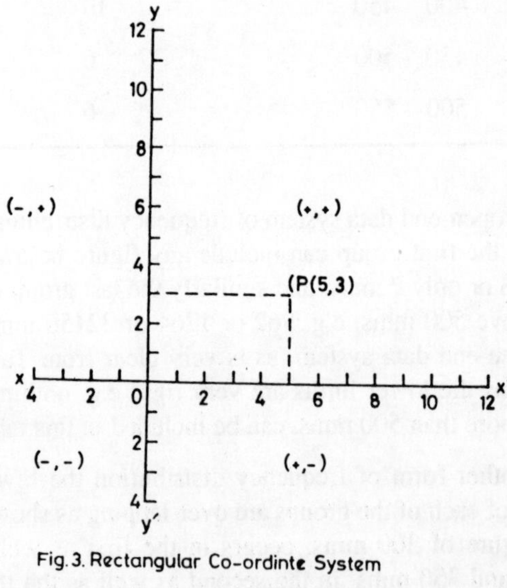

Fig. 3. Rectangular Co-ordinte System

TABLE X

Annual Rainfall (mms.)

Year	Rainfall
1941	310
1942	290
1943	280
1944	300
1945	320
1946	310
1947	340
1948	280
1949	290
1950	300

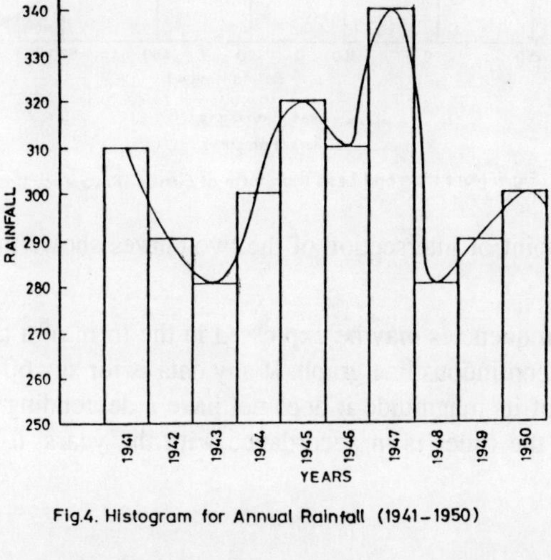

Fig.4. Histogram for Annual Rainfall (1941–1950)

The type of frequency distribution shown in Figure 4 is known as Histogram. The same frequencies can be shown by frequency curve which can be drawn by joining the middle points at the top of each rectangle as is shown by a curved line in Figure 4. The advantage of such a curve is that it provides a visual representation and the pattern that the area would follow. This idea cannot be conveyed by the data alone put in tabular form.

Graphic frequency can be combined with tabular frequency also. If the curves are drawn for more than and less than tables with the help of Tables VI and VII, they will assume a pattern shown in Figure 5.

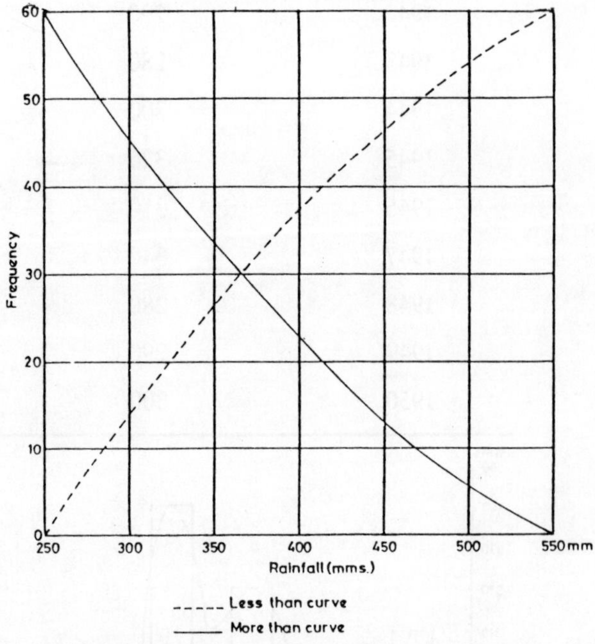

Fig.5. More than and Less than form of Curves (Frequency Basis)

The point of intersection of the two curves shows the position of normal.

The frequencies may be expressed in the form of a table or in the form of a continuous line graph. If any data is for say 60 years and on the basis of its magnitude it does not have a descending or ascending order but the order is in accordance with the years, it is said to be having a temporal orderly record. It should therefore first be arranged on the basis of its magnitude and then groups formed in over-lapping or non-over-lapping form.

There is yet another system of gradation as shown in Table XI which is although free from all ambiguities, one does not come across this system as far as geography is concerned.

TABLE XI

Frequency Distribution (Under Form)

Grades (mms.)	Frequency
250 under 300	15
300 under 350	11
350 under 400	12
400 under 450	10
450 under 500	6
500 under 550	6

To draw a graph with the help of the graded data shown in Table XI a table can be prepared as shown in Table XII.

TABLE XII

Frequency Distribution (Less Than Form) (Proportions)

Grades	Frequency	Rainfall (mms.)	Frequency (Less Than)	Proportion
250 – 300	15	250	0	0/60 = 0.00
300 – 350	11	300	15	15/60 = 0.25

350 – 400	12	350	26	26/60 = 0.43
400 – 450	10	400	38	38/60 = 0.63
450 – 500	6	450	48	48/60 = 0.80
500 – 550	6	500	54	54/60 = 0.90
550 – 600	0	550	60	60/60 = 1.00

TABLE XIII

Frequency Distribution (More Than Form) Proportions

Grade	Frequency	Rainfall (mms.)	Frequency (More than)	Proportion
250 – 300	15	250	60	60/60 = 1.00
300 – 350	11	300	45	45/60 = 0.75
350 – 400	12	350	34	34/60 = 0.57
400 – 450	10	400	22	22/60 = 0.37
450 – 500	6	450	12	12/60 = 0.20
500 – 550	6	500	6	6/60 = 0.10
550 – 600	0	550	0	0/60 = 0.00

With the help of the proportions worked out in Tables XII and XIII, a graph may be plotted as shown in Figure 6.

The point of intersection of the two curves defines the mean. The 'Less Than' curve shows a gradual increase whereas the 'More Than' curve shows a gradual decrease.

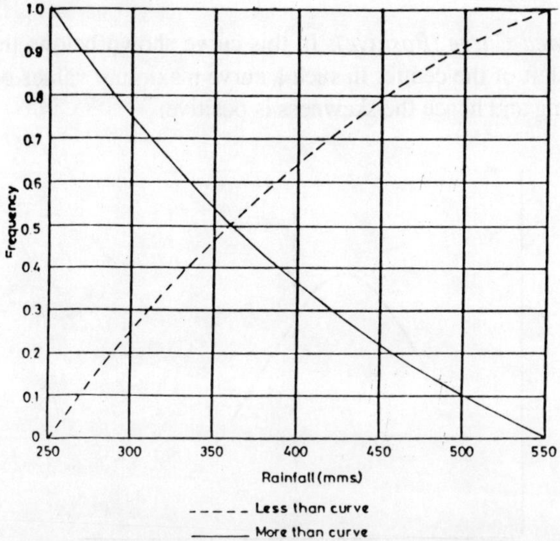

Fig. 6 More than and Less than form of Curves (Proportional Basis)

Typical Patterns of Frequency Distribution

Varied types of data when plotted bring out typical patterns of frequency curves some of which may be classified as follows:

a) Curve of Constancy: The curve shown below is actually a straight line indicating that the position of the variates shown does not have any change year after year.

b) Normal Curve: A curve is said to be a normal curve when the rate of variation on either side of the peak is the same as shown below. It is supposed to be the most basic curve of all statistical analysis. It serves as 'Normal'.

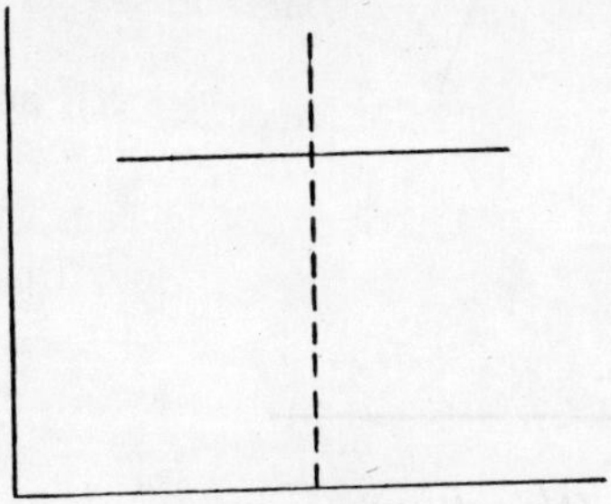

Constancy Curve

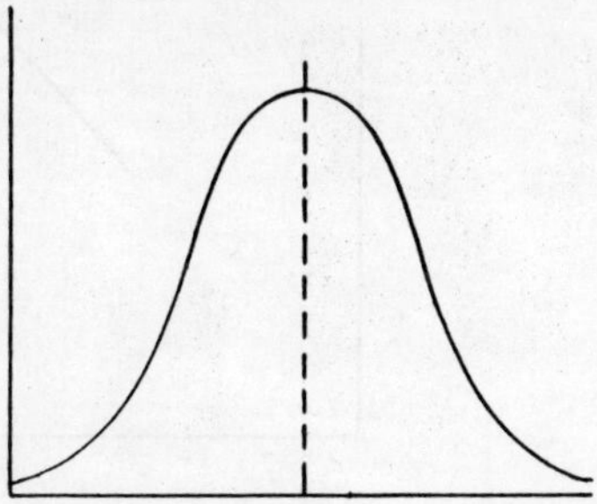

Normal Curve

c) Skewed Curve (Positive): In this curve shown below, the peak falls to the left of the centre. In such a curve maximum values occur in the beginning and hence the skewness is positive.

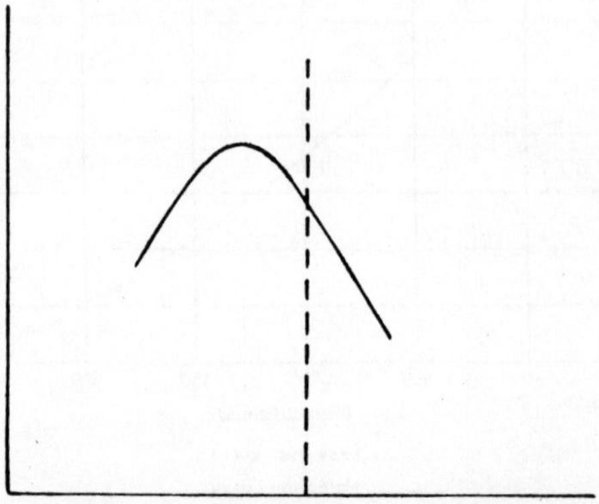

Skewed Curve (Positive)

d) Skewed Curve (Negative): The only difference in this and the preceding curve is that the peak falls to the right of the centre showing that maximum values occur later as shown below. Such a skewness is negative.

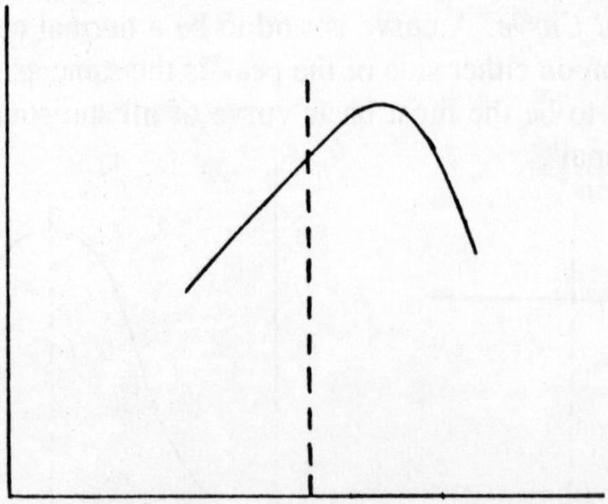

Skewed Curve (Negative)

e) Multi-modal Curve: Such a curve has two peaks. One to the left of the centre and the other to the right as shown below. Such a curve has high degree of variability.

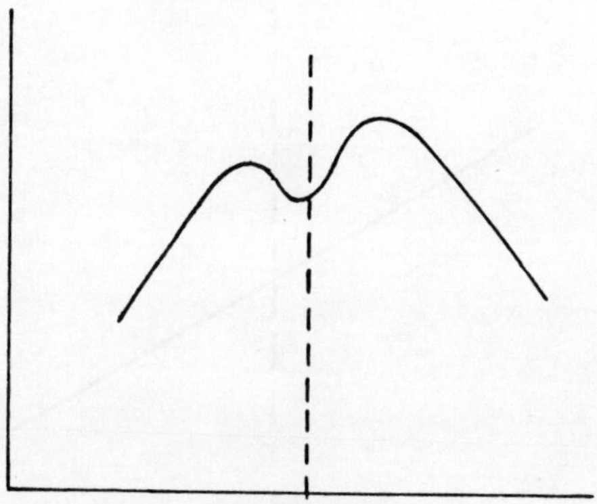

Multimodal Curve

f) Constantly Increasing Curve: Such a curve shows that the values are increasing at the same rate as shown below.

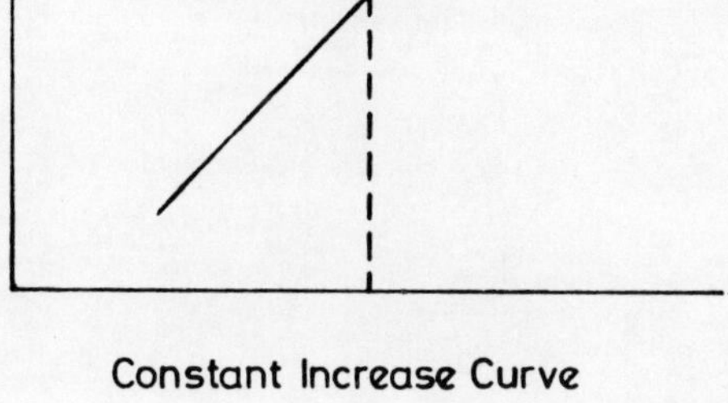

Constant Increase Curve

g) Constantly Decreasing Curve : Just opposite to the previous one, the values in it are decreasing at the same rate as indicated below.

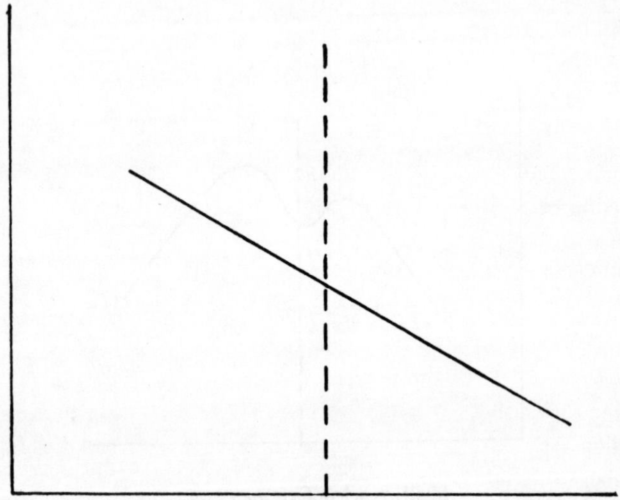

Constant Decrease Curve

CHAPTER III

MEASURES OF CENTRAL TENDENCY

Average is a measure of location or position of frequency distribution on the scale of variates i.e. a single value which is representative of the entire population. Average is of a very great significance as far as comparisions are concerned.

Let there be five items 4 5 6 7 and 8 arranged in ascending order. In these items 6 is the figure which has two items below it and two above it i.e. 6 is a central figure that divides the data in two halves having 6 as a particular location. The figure of 6 in this case is called Median. The total of the five figures which is 30 when divided by the number of items i.e. 5 gives 6 which is the average but technically it is called Arithmetic Mean. In the data 2 5 5 5 6 7, there are six items in which the frequency of 2 is 1, of 5 is 3 and of 6 and 7 is 1. By simple inspection, 5 looks typical because it appears thrice. The figure of 5 in this case is called Mode which also is a particular measure of location and yet another type of average. The commonly used averages are of five types:

Arithmetic Mean

Median

Mode

Geometric Mean and

Harmonic Mean

ARITHMETIC MEAN

Averages are used as position of reference for objective interpretation of frequency distribution and they are used specially for the computation of the techniques of variability in any frequency distribution. Arithmetic mean is tne simplest form of various types of averages which is defined as the sum of the items in the data divided by the number of items in the data. In other words it would be the share of each individual value in operation if the values are equally distributed. In the following 11 figures

68 68 67 67 67 66 65 65 65 64 64

The total is 726. Hence the arithmetic mean would be 726/11 = 66. For generalized formula, suppose the figures are indicated as x_1 x_2 x_3 x_4 ....x_n, the arithmetic mean would be

$$(x_1+x_2+x_3+x_4+..........x_n)/n$$

The genralized formula would be

$\bar{x} = \dfrac{\sum_{i=1}^{n} x_i}{n}$ where $\bar{x}$ represents the arithmetic mean, $\sum_{i=1}^{n} x_i$, the total of individual values (i) in which the value of (i) ranges from 1 to n and n is the number of items in the data. If the record is for 50 years the formula could be interpreted thus

$$\bar{x} = \sum_{i=1}^{50} x_{50} \; /50$$

While calculating arithmetic mean, summation is a must. When the data is for a small duration, the summation is very easy but if it is for a very long period say 500 years, the summation is quite a tedious job which is unavoidable. Therefore to make the job easier, the following methods of summation have been introduced:

a) Ordinary System of Summation,

b) Different Origin System of Summation and

c) Mid-point System of Summation.

a) Ordinary System of Summation

Suppose the data for 11 years of rainfall in millimetres for any place A is as follows

64 64 65 65 65 66 67 68 68 69 69

The data can be directly added.

The above data can be written in frequency form also and added thus:

Summation (Ordinary System with Frequencies Given)

Values	Frequency	Total Value
64	2	64 × 2 = 128
65	3	65 × 3 = 195
66	1	66 × 1 = 66
67	1	67 × 1 = 67
68	2	68 × 2 = 136
69	2	69 × 2 = 138
Total	11	730

i.e. $\overline{x}$ = 730/11 = 66.36

In the above case the summation for finding out the arithmetic mean is quite easy.

b) Different Origin System of Summation

All the arithmetic numbers have there origin as zero. 1 means 0 + 1, 2 means 0 + 2 and so on. Using the preceding table and taking 60 as origin, 64 can be written as 60 + 4 and the summation can be done as follows:

Summation (Different Origin System)

Values	Deviation from Origin (60)	Frequency	Deviation × Frequency
64	4	2	4 × 2 = 8
65	5	3	5 × 3 = 15
66	6	1	6 × 1 = 6
67	7	1	7 × 1 = 7

68	8	2	$8 \times 2 = 16$
69	9	2	$9 \times 2 = 18$
Total		11	70

i.e. $\bar{x} = 70/11 = 6.36$ and therefore the actual arithmetic mean is 60 + 6.36 = 66.36 (60 being the origin). There is no hard and fast rule as far as the selection of origin is concerned but it is always advisable to have it around the actual data.

c) Mid-point System of Summation

Before taking up Mid-point System of Summation it would be advisable to know about different types of grouped data. Basically there are two types of grouped data—the over-lapping and the non-over-lapping types. When the data is in over-lapping form, the mid-point has to be calculated. The calculation is done by adding the upper and the lower limits of each group and then dividing by 2.

Illustration:

Groups
10 – 15
15 – 20
20 – 25

The mid-point in the first group will be

$$(10 + 15)/2 = 25/2 = 12.5$$

The mid-point in the second group will be

$$(15 + 20)/2 = 35/2 = 17.5 \text{ and so on.}$$

On the other hand if the grouped data is in non-over-lapping form as shown below:

Groups
10 – 14
15 – 19
20 – 24

and so on, it is said to be in discrete form. Such a grouped data therefore has to be converted first into continuous form by subtracting ·5 from the lower limit and adding .5 to the upper limit of the same group. The figure of .5 in this case has been obtained by dividing the difference between the upper limit of one group and the lower limit of the following group by 2. The new group formation will be as follows:

Groups

9.5 – 14.5

14.5 – 19.5

19.5 – 24.5 and so on.

The mid-point will now be calculated as described earlier.

The difference between the upper limits of two consecutive groups gives the class interval or the widths of the groups irrespective of the fact that the data is in over-lapping or non-over-lapping grouped form.

When the data is in grouped form the summation becomes somewhat difficult. In this case since the actual figures are not known exact mean can also not be calculated because exact total cannot be found out. The values of the items in each of the groups are evenly distributed therefore the middle point of each group can be found out and that would be taken as the average for that group and can be calculated as shown in Table XIV.

TABLE XIV

Summation (Mid-point System)

Groups	Mid-point value	Frequency (f)	Mid-point × f
63 – 65	64	2	64 × 2 = 128
65 – 67	66	4	66 × 4 = 264
67 – 69	68	3	68 × 3 = 204
69 – 71	70	2	70 × 2 = 140
			Total = 736

Therefore the arithmetic mean $\overline{x} = 736/11 = 66.91$

The mean calculated earlier with the help of the individual values when each of the items were known was 66.36. But in the grouped data prepared with the help of the same table, the mean comes as 66.91 i.e. a difference of .55 which is almost negligible. Meaning thereby that the mean can be calculated even if the data is in grouped form. But it should be noted that this proposition holds good only when the data is in 'Close-end' form. If the data is in 'Open-end' form, the mean cannot be calculated because the mid-point value of the first and the last group cannot be found out. The mean when the data is in grouped form calculated earlier can be put in generalized formula as follows:

$$\overline{x} = \sum_{j=1}^{n} f_j x_j / n$$

Where f_j is the frequency

x_j is the mid-point value and

n is the number of records in the data.

A still shorter cut for calculating the arithmetic mean would be by selecting any assumed mean.

Illustration:

60 64 65 65 65 66 67 68 68 69 69

Take assumed mean as 65 and find the deviation from the assumed mean.

x_i	Deviation from Assumed Mean (65)
60	60 – 65 = –5
64	64 – 65 = –1
65	65 – 65 = 0
65	65 – 65 = 0
65	65 – 65 = 0

66	66 – 65 = +1
67	67 – 65 = +2
68	68 – 65 = +3
68	68 – 65 = +3
69	69 – 65 = +4
69	69 – 65 = +4
	Total =+11

$$\bar{x} = 65 \text{ (assumed mean)} + 11/11$$

$$= 65 + 1 = 66$$

The formula for which would be

$\bar{x} = a + \Sigma d/n$ where 'a' is the assumed mean,

'd' is the deviation from the assumed mean and

'n' is the number of records.

For the grouped data also the same method can be applied. Let the assumed mean in the following grouped data be 65.

Groups	Frequency (f)	Mid-point value (x_j)	Deviation from Assumed Mean (65) ($d = x_j - a$)	$f \times d$
60 – 62	1	61	61 – 65 = – 4	1 × – 4 = – 4
62 – 64	0	63	63 – 65 = – 2	0 × – 2 = 0
64 – 66	4	65	65 – 65 = 0	4 × 0 = 0
66 – 68	2	67	67 – 65 = + 2	2 × + 2 = + 4
68 – 70	4	69	69 – 65 = + 4	4 × +4 = +16
				Total = +16

Arithmetic Mean is therefore

$$= 65 \text{ (assumed mean)} + 16/11$$

$$= 65 + 1.4$$

$$= 66.4$$

Mathematically it can be expressed as

$$\bar{x} = a + \Sigma fd/n$$

Characteristics of Arithemetic mean:

1. It cannot be located by inspection alone.

2. To find the arithmetic mean, the data may or may not be arranged in ascending or decending order.

3. The sum of the deviations of the individual values from the arithmetic mean is always zero.

Illustration:

Individual Values	60	64	65	65	65	66	67	68	68	69	69
Arithmetic Mean	66	66	66	66	66	66	66	66	66	66	66
Deviation from Arithmetic Mean	–6	–2	–1	–1	–1	0	+1	+2	+2	+3	+3

The sum total is (–11) + (+11) = 0

Mathematical Proof:

$(x_1 - \bar{x}) + (x_2 - \bar{x}) + (x_3 - \bar{x}) + ... + (x_n - \bar{x})$ should be equal to zero.

or $(x_1 + x_2 + x_3 + ... + x_n) - (\bar{x} + \bar{x} + \bar{x} + \text{.. n times})$

or $(x_1 + x_2 + x_3 + ... + x_n) - n\bar{x}$

dividing both sides by n we have

$[(x_1 + x_2 + x_3 + ... + x_n) / n] - n\bar{x} / n$

or $(\bar{x} - \bar{x})$ which is equal to zero.

4. The sum of the squares of the deviations of individual values from the arithmetic mean is minimum i.e. always less than the sums of the squares of the deviations calculated with respect to Median or Mode.

Illustration:

Items	Deviation from Mean (66)	Squares of Deviation from Mean	Deviation from Mode (65)	Squares of Deviation from Mode	Deviation from Median (67)	Squares of Deviation from Median
54	–12	144	–11	121	–13	169
65	–1	1	0	0	–2	4
65	–1	1	0	0	–2	4
65	–1	1	0	0	–2	4
66	0	0	+1	1	–1	1
67	+1	1	+2	4	0	0
68	+2	4	+3	9	+1	1
68	+2	4	+3	9	+1	1
69	+3	9	+4	16	+2	4
69	+3	9	+4	16	+2	4
70	+4	16	+5	25	+3	9
	Total	190		201		201

In all the three totals calculated with respect to mean, median and mode, the value given by mean (190) is the least. The mathematical proof of the fourth characteristic can also be given but since it requires a good understanding of mathematics, it has been avoided.

5. Every individual figure enters into the arithmetic mean.

6. In an arranged data, arithmetic mean is always biased towards extreme values.

7. Arithmetic mean completely conceals the unevenness of the data.

MEDIAN

Median is the central value or the median value which divides the number of items in a list into two halves such that half of the values are above and half below the median value. Median is also a measure of certrality in which the number of items are counted but the magnitude is not involved. The arrangement of values is in order of magnitude–ascending or descending. Medians are also defined as fiftieth percentile. If n is the number of items, the median would be $\frac{(n+1)}{2}$ th item.

Median can also be calculated when the data is in grouped form the formula for which is

$$M_d = L_1 + \frac{W(n/2 - C)}{f}$$

Where M_d is the median when the data is in grouped form,

L_1 is the lower limit of the group in which the median is expected,

W is the width or the class interval,

n is the total of frequencies,

C is the total of frequencies preceding to that group in which the median is expected and

f is the frequency in that group in which the median is expected.

Illustration:

Group	Frequency
62.5 – 64.5	2
64.5 – 66.5	2

66.5 – 68.5	4
68.5 – 70.5	3

Total frequency is 11. It means that the median would be (11 + 1)/2 = 12/ 2 = 6th item.

In the first group the frequency is 2 so there is no question of median lying in this group. The cumulative frequency of the first and the second group is 2 + 2 = 4, so the second group also does not have the median. The cumulative frequency of the first, second and the third groups is 2 + 2 + 4 = 8 i.e. the 6th item lies in this group so that the median value lies in the third group. Now substituting the values in the formula, we get

$$M_d = 66.5 + \frac{2(11/2 - 4)}{4}$$

$$= 66.5 + \frac{2(5.5 - 4)}{4}$$

$$= 66.5 + 3/4$$

$$= 66.5 + .75$$

$$= 67.25$$

The result thus obtained can be verified with the help of another formula:

$$M_d = L_2 + \frac{W(C - n/2)}{f}$$

where L_2 is the upper limit of the group in which the median is expected,
C is the total of frequency of the groups following that group in which the median is expected,
W is the class interval,
n is the number of the items and
f is the frequency in that group in which the median is expected.

Substituting the values and using the same data we get

$$M_d = 68.5 + \frac{2(3 - 11/2)}{4}$$

$$= 68.5 + \frac{2(3 - 5.5)}{4}$$

$$= 68.5 + \frac{2(-2.5)}{4}$$

$$= 68.5 - \frac{5}{4}$$

$$= 68.5 - 1.25$$

$= 67.25$ which is correct.

In an open end data distribution, the mean cannot be calculated but since the values of occurences are defined, median can be calculated.

Characteristics of Median:

1. Median is the 50th percentile which balances the entire data.

2. In median, the values of individual items are not taken into consideration.

3. Median is not influenced by the extreme values and variations.

4. It can be located by inspection alone.

5. For the calculation of median, the data is to be arranged in ascending or descending order.

6. Median is not very suitable for mathematical calculations.

7. Median is very much biased towards extreme values.

MODE

Mode literally means fashion i.e. the style followed by the majority. Mode in statistics is that value of the item which occurs more than once in the distribution. It is the point of greatest density. In the data 2 3 4 5 5 5 5 6 9, the mode is 5 because it occurs maximum number of times. The data is therefore uni-modal. The distribution may be bi-modal if the data has two modes e.g.

2 3 3 4 5 6 6 6 7

In the above data there are two distinct peaks. In the same way a data is tri-modal if it has three modes and if there are more than three modes, the data is said to be multi-modal.

The location of mode is very simple and no formulation is required but when there is a grouped data the task becomes somewhat difficult and formulation is required.

Before working out the calculations for mode in the grouped data, certain points should be remembered.

1. The mode occurs in that group which has the maximum frequency.

2. The group which has the maximum frequency should be named f and the group prior to that, f_1 and the one following f should be named f_2.

In the grouped data the formula for calculating the mode is as follows:

$$\text{Mode} = L_1 + \frac{W(f - f_1)}{(f - f_1) + (f - f_2)}$$

$$\text{or} = L_1 + \frac{W(f - f_1)}{2f - f_1 - f_2}$$

Where L_1 is the lower limit of the group in which the mode is expected

W is the width or the class interval.

Illustration:

Groups	Frequency	
Below 60	8	
60 – 64	15	f_1
64 – 68	40	f
68 – 72	38	f_2
72 – 76	10	
over 76	5	

Substituting the values in the formula, we get

$$\text{Mode} = 64 + \frac{4(40 - 15)}{(40 - 15) + (40 - 38)}$$

$$= 64 + \frac{4(25)}{25 + 2}$$

$$= 64 + 100/27$$

$$= 64 + 3.7$$

$$= 67.7$$

The mode is only a theoritical deduction which is not much used.

Characteristics of Mode:

1. It can be located by inspection.

2. There is no influence of extreme values on the mode.

GEOMETRIC MEAN

As for as geography is concerned, Geometric Mean is utilised in computing the rate of increase generally.

Illustration:

Let the increase in population in two years be 44 per cent.

With this data alone, yearly rate of increase of population can be found out.

Let the increase of population in one year be 22 per cent.

If the population at the beginning of first year is 100.

The population at the beginning of second year would be 122.

Therefore with 22 per cent rate of increase, the increase in population at the beginning of third year would be $\frac{22 \times 122}{100}$ = 2684/100 = 26.84.

Therefore at the beginning of the third year the population would be 122+26.84 = 148.84. Here the increase comes as 48.84 per cent instead of 44 per cent. It means that arithmetic mean is not required but geometric mean has to be taken into consideration which is the root of the products of the items i.e.

$$\text{Geometric Mean} = \sqrt[n]{x_1 \times x_2 \times x_3 \times x_4 \times \ldots x_n}$$

Suppose there are two items 3 and 27.

The geometric mean would be $= \sqrt{3 \times 27} = \sqrt{81} = 9$

If there are suppose 6 items then the 6th root of the product of the six items will have to be taken.

In the above exercise, with 44 per cent increase in population in two years, the population would become 144 if the initial population is taken as 100. The rate of increase would therefore be the geometric mean of 100 and 144.

The increase is therefore $= \sqrt{100 \times 144} = 10 \times 12 = 120.$

Therefore the average rate of increase would be 120 – 100 = 20 per cent per year.

HARMONIC MEAN

Harmonic Mean is the sum of the reciprocal of the arithmetic mean of the reciprocal of the variates in the distribution.

If the variates are $x_1\ x_2\ x_3 ....\ x_n$, the reciprocal would be

$$1/x_1 + 1/x_2 + 1/x_3 + ...+ 1/x_n \quad \text{(a)}$$

The arithmetic mean would therefore be

(a)/n.

The harmonic mean would be 1/(a)/n or n/(a).

The harmonic mean is not much used in geography and it is important only in those cases where gradient is involved.

CHAPTER IV

MEASURES OF DISPERSION

Averages do not give a clear idea regarding the distribution of data in any particular aspect e.g. if there are 60 persons with the arithmetic mean of their ages as 35, the median as 40 and the mode as 37 years. From the arithmetic mean one could know only that 35 years of age may be out of 5000 persons. Similarly the median of 40 indicates that 50 per cent population has ages above 40 and 50 per cent below 40 years and in the same way mode of 37 years of age would only show that most of the persons in the data were 37 year old. The main thing to be found out is as to how much the average is typical and how much the data varies in size. For this purpose, Dispersion, Scatter, Variation or Variability are a measure of extent to which the individual items vary from a particular average.

The various means to determine these aspects are as follows:

a) Range

b) Quartile Deviation

c) Mean Deviation and

d) Standard Deviation

a) RANGE

Range is the simplest form of measure of dispersion. It is the difference between the maximum and the minimum value in any data. Since the range depends upon extreme values in any data, it is not a very appropriate measure of dispersion.

b) QUARTILE DEVIATION

Quartile deviation is also known as semi inter quartile range. Quartile is a specific location in a data when it is arranged in ascending or descending order. It has two sub-types. Upper Quartile and Lower Quartile. Upper Quartile is that location in a data which will have about 75 per cent items below that value and 25 per cent above that.

Similarly Lower Quartile is that value above which are about 75 per cent items and below which 25 per cent when the data is arranged in ascending or descending order.

In the following figures the Upper and the Lower quartiles can be found by inspection as well as with the help of a simple formula:

3 9 21 53 74 101 110

The formulation is required when the items in the data are very large. To calculate the Lower Quartile, the formula would be

Q_1 (Lower Quartile) = $\frac{1}{4}$(n+1)th item in the data e.g. in the above data, it would be $\frac{1}{4}(7+1)$ = 2nd item.

Similarly the Upper Quartile can be calculated with help of the following formula:

Q_u (Upper Quartile) = $\frac{3}{4}$ (n+1)th item i.e. in the above data

$$Q_u = \frac{3}{4} \times (7+1)$$

$$= \frac{3}{4} \times 8 = 24/4 = 6\text{th item.}$$

In the above formula, n stands for number of items in the data.

Sometimes when the number of items in the data are even there arises some problem but the same formula is used.

Illustration:

Let the figures be as follows:

3 11 42 59 64 68 73 79 80 92 97 99

Substituting the values in the formula we have

$$Q_1 = \frac{1}{4}(12 + 1) = 13/4 = 3.25\text{th item in the data.}$$

i.e. Q_1 will be the third item (42) plus .25th item after that. .25th item in this case would be calcuated by multiplying .25 with the difference between the 3rd and the 4th item i.e. .25 (59 – 42) = .25 × 17 = 4.25.

Hence the lower quartile will be 42+4.25 = 46.25.

$$\text{Similarly } Q_u = \frac{3}{4}(12+1) = \frac{3}{4} \times 13 = 39/4 = 9.75.$$

i.e. 9th item (80) plus .75th item.

i.e. $.75\,(92 - 80) = .75 \times 12 = 9$

$$\text{Hence } Q_u = 80 + 9 = 89$$

In the same way octile can be calculated using the formula

$$0_1 \text{ (First Octile)} = \frac{1}{8} \times (n+1)\text{th item}$$

$$0_2 \text{(Second Octile)} = \frac{2}{8} \times (n+1)\text{th item and so on.}$$

And in the same way deciles can be calculated using the following formula:

$$D_1 \text{ (First Decile)} = \frac{1}{10} \times (n+1)\text{th item}$$

$$D_2 \text{ (Second Decile)} = \frac{2}{10} \times (n+1)\text{th item and so on.}$$

Inter quartile ranges and quartile deviations are very important in the study of several geographical studies. They are worked out thus:

Let the rainfall for 15 years be as follows:

11 19 23 26 29 34 39 47 53 69 69 75 79 87 94

$$Q_1 = \frac{1}{4} \times (15+1) = 16/4\text{th item i.e. 4th item which is 26}$$

$$Q_u = \frac{3}{4} \times (15+1) = 48/4 = 12\text{th item i.e. 75.}$$

The inter quartile range would be

$$75 - 26 = 49$$

The quartile deviation is found by dividing the inter quartile range by 2 i.e.

$$\text{Quartile Deviation } (Q_u - Q_1)/2 = (75 - 26)/2 = 49/2 = 24.5.$$

The quartile deviation is a better measure of dispersion but the handicap is that it does not take into account every individual value in the data. Inter quartile range provides an index of variability in 50 per cent of the total values. If the distribution is perfectly symmetrical, the range should be such that half of it indicates distance of quartile from the median.

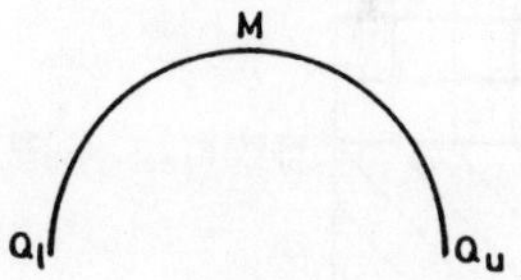

If Q_u and Q_1 are equidistant from the median, then the distribution will be symmetrical but if the distribution is asymmetrical, there will be a greater bias towards Q_u and lesser bias towards Q_1 as shown in the above figure.

The use of quartile technique in the preparation of choropleth maps is one of the latest techniques in graphic representation of geographic data. Grading of raw data in groups for graphic representation is very often quite a perplexing problem for the geographers. The groupings do not obey any hard and fast rule. It is done just arbitrarily. The quartile technique of grouping cannot be done arbitrarily. It does follow certain principles e.g. the data is first of all arranged in simple array in which the median, upper quartile, the lower quartile and the upper and the lower limits of the data are located. In this technique, howsoever long the data may be, the choropleth map will always have four shades. The darkest shade at the top most box and the lightest shade at the bottom most box.

Illustration:

Let the median be 35, the upper quartile 98.5, the lower quartile 8, the upper limit i.e. the highest figure in the data as 127 and the lower limit i.e. the lowest figure in the data as 5. The index would be prepared as shown on the next page.

The figures falling in various categories are given the corresponding shades while preparing choropleth maps.

Another very useful aspect of quartile technique is the preparation of dispersion diagrams. Arithmetic means do relate to the whole body of the data which are single values and which are no doubt very sound mathematically but they may not show the variations in the data. Take

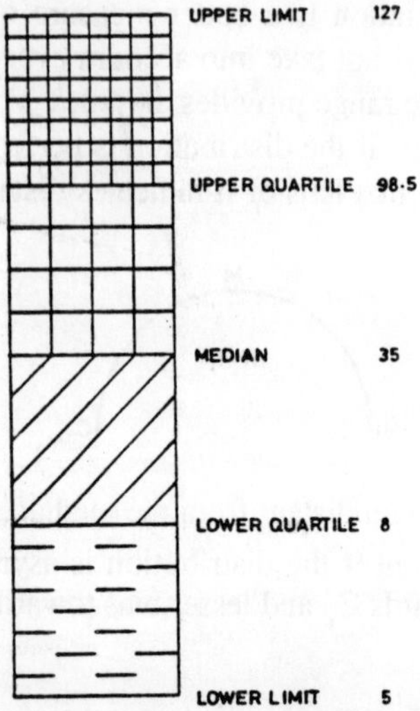

for example the following two sets A and B:

A	B
25	23
23	24
20	18
26	25
18	26
Total 112	116

The arithmetic mean of A is 112/5 = 22.4

The arithmetic mean of B is 116/5 = 23.2

The arithmetic means of A and B do not indicate whether there is significant variation in the data from the mean or not. To know the degree of variation, parametric or non-parametric tests may help a great deal but a graphic form of representation could provide a much quicker and better technique of comparison. The comparison can be done by line graph or bar graph also but it would not bring out a clear picture. Thc dispersion diagram provides a very clear visual picture of comparison.

For making a dispersion diagram, the data is arranged in simple array and divided in sections e.g. quartiles and graphically represented in the form of a bar diagram and individual values plotted by dots.

Mathematically only the arithmetic mean is the most sound in all types of averages but median has other advantages. Median also does not give any idea about the distribution of individual values or the magnitude of the data. For this purpose the values are dispersed in slabs. Median is the 50th percentile which is divided into two halves– Upper and Lower quartiles which are 25th and 75th percentiles respectively. Quartiles may further be sub-divided into octiles. By plotting the median, quartiles, no doubt one gets an idea of comparisons but when all the individual values are plotted, the picture is still more clear. Care should be taken that the values plotted should be of not less than 30 records. When greater precision is to be taken into account, the data should be of not less than 50 records. In geographical studies, the preparation of dispersion diagram (Fig. 7) specially to show a variation in rainfall etc. is very helpful. In a country like India where agriculture is the mainstay of its economy, rainfall variations specially during the growing season have to be very precisely understood because the variation in amount and time may seriously tell upon the agricultural production.

The diagram after the data of stations A and B below does indicate variations but if some specific cases are considered, it would be much easier to understand.

	A	B	
	26	27	
Q_u	26	25	Q_u
	24	23	
M	23	23	M
	21	21	
Q_l	20	18	Q_l
	18	17	

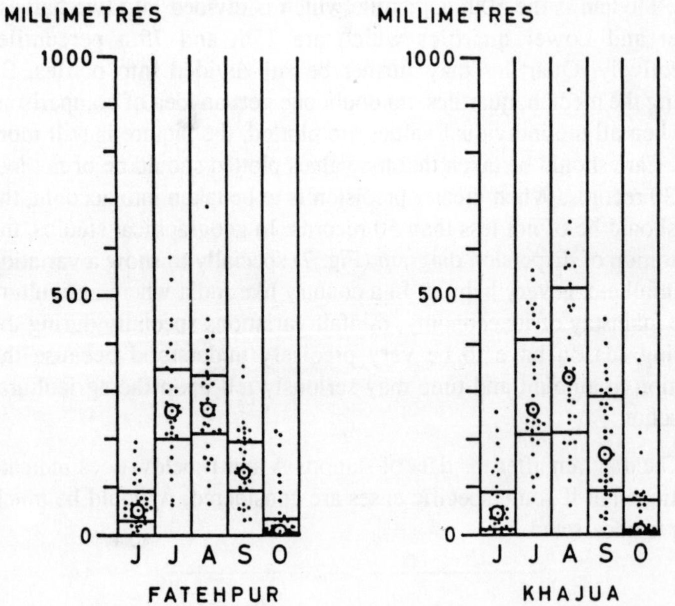

RAINFALL DISPERSION DIAGRAMS

JUNE-OCTOBER

1935-1969

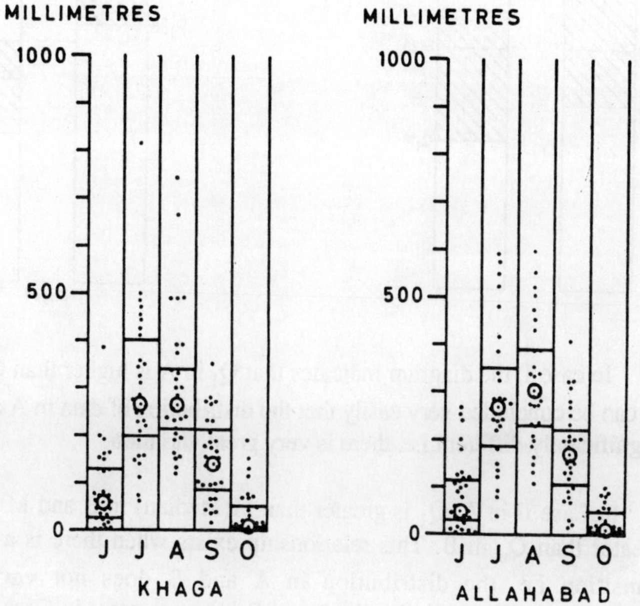

RAINFALL DISPERSION DIAGRAMS

JUNE-OCTOBER

1935-1969

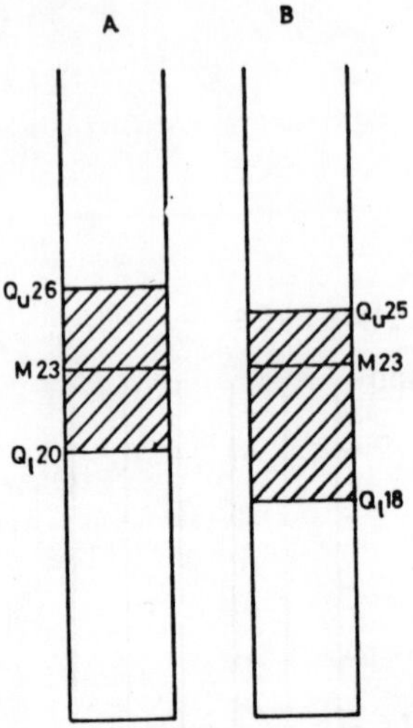

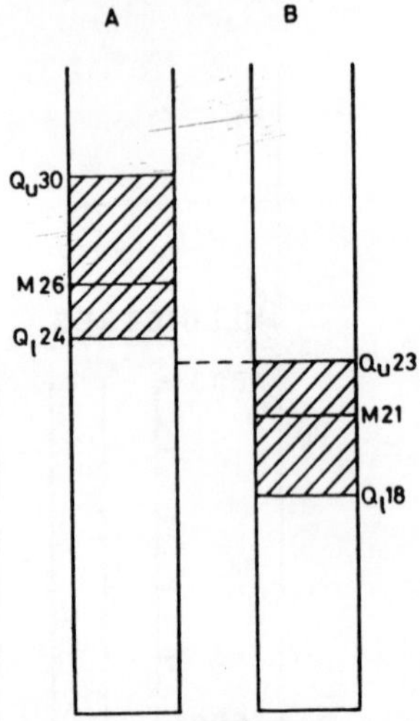

Case I

In case I, the diagram indicates that Q_l in A is higher than Q_u in B. It can be concluded verv easily that the distribution of data in A and B is significantly different i.e. there is very great variation.

In Case II in A, Q_l is greater than M(Median) in B and M in A is greater than Q_u in B. This relationship exists when there is a sort of transition i.e. the distribution in A and B does not vary very significantly. The variation in A and B is not as great in Case II as it was in Case I.

In Case II we had

Q_l in A greater than M in B (a)

M in A greater than Q_u in B.

In Case III, Q_l in A greater than M in B (b)

and M in A lesser than Q_u in B.

In (a) of Case II and (b) of Case III the conditions are the same i.e. there can be pairs of conditions.

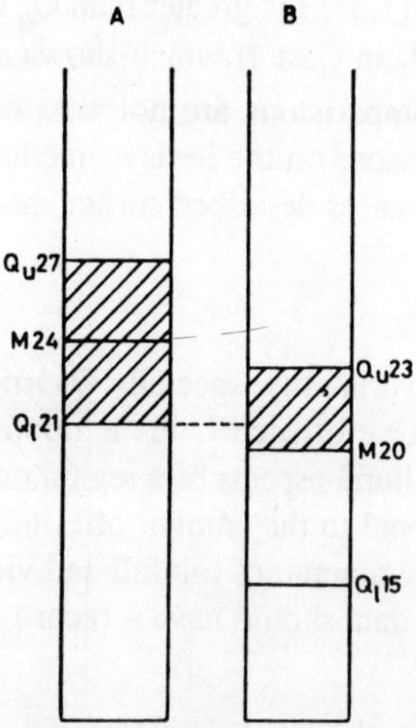

Case II

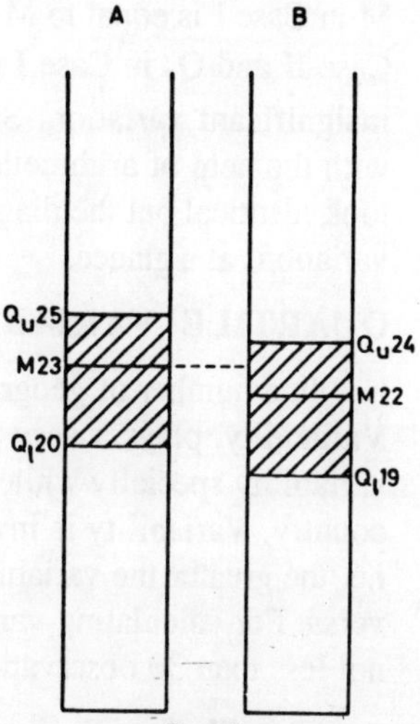

Case III

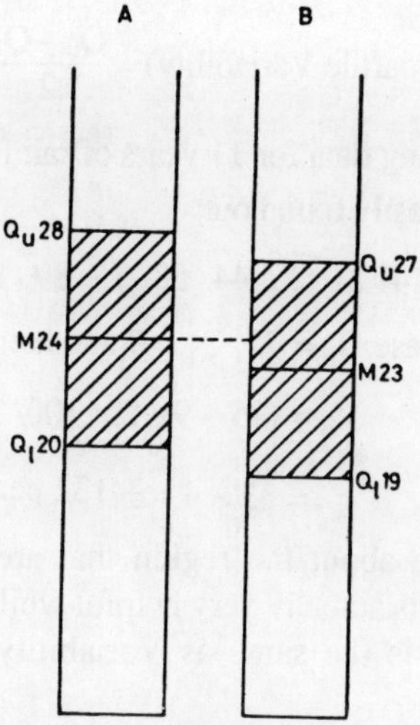

Case IV

In Case IV, Q_l in A is not greater than Q_u of B or M of B.

In cases II and III, the distribution is of a different type but not significant. But in Case II, the variation is slightly greater but not as great as in Case III. In Case II, there is no noticeable difference. There may be a number of other cases resembling those described earlier e.g.

M in Case I is equal to M in Case II; Q_u in Case I is greater than Q_u in Case II and Q_1 in Case I is greater than Q_1 in Case II which shows an insignificant variation. Such types of comparisions are not possible with the help of arithmetic mean. Comparisons on the basis of median look identical but the diagrams in the four cases described earlier show variations at a glance.

QUARTILE VARIABILITY

In a number of geographical studies, variability, specially Quartile Variability, plays a very important part. One of such fields is rainfall variability specially while studying agricultural aspects of a region or a country. Variability is inversely proportional to the amount of rainfall i.e. the greater the variability the lesser the amount of rainfall and vice versa. For calculating variability also the data should have a record of not less than 30 observations.

Quartile variability is expressed as the quartile deviation as percentage to the median.

$$V_q \text{ (Quartile Variability)} = \frac{Q_u - Q_1}{2} \times \frac{100}{\text{Median}}$$

From the following data for 11 years of rainfall record, Q_u , Q_1 and median can be very easily found out:

3 7 9 14 21 25 44 58 75 89 104

In this case $Q_u = 75$, $Q_1 = 9$ and median = 25.

$$V_q = (75 - 9)/ 2 \times 100/ 25$$

$$= 33 \times 4 = \pm 132 \text{ per cent.}$$

It gives an idea about the region that are prone to drought or excessive rainfall. It is actually very helpful while planning agriculture. Quartile variability is the same as Variability Index or Variability Percentage.

MEAN DEVIATION

Frequency distribution curves provide simplest and the most effective indicator of scatter. The dispersion diagram in which every individual value is plotted is also quite helpful. Scatter can roughly be expressed in terms of highest and lowest values observed in the data. The following two sets of data have the same mean as 6:

Set I	Set II
4	2
5	3
6	4
7	5
8	16
Total 30	Total 30

$\bar{x} = 30/5 = 6$ $\bar{x} = 30/5 = 6$

To use this arithmetic mean of 6 more effectively so as to express the entire data, maximum and minimum values which are 4 and 8 in Set 1 and 2 and 16 in Set II are taken. The range in Set I is 4 to 8 and the mean is 6 whereas in Set II the range is 2 to 16 and the mean is the same as 6. In Set I the difference between mean and minimum and maximum value is 2 whereas in Set II, the difference from mean of minimum value is 4 and of maximum value is 10 i.e. in Set I, the variates are nearer to the mean or that the data occupies near central position from the mean i.e. the data is more compact. On the other hand in Set II the data is not so compact about the mean but it is more scattered. It is therefore concluded that the longer the range the lesser the compactness and greater the scatter i.e. upto some extent the range can help the compactness of the data but it is highly a generalized expression which does not convey very specific information about the individual values of the data.

The difference of individual values from the mean is called deviation. The sum of the deviation of individual values from the mean is called total deviation. The total deviation when divided by the number of items in the data is called Mean Deviation which is another measure of dispersion expressed as

$$\text{Mean Deviation} = \frac{1}{n}\sum_{i=1}^{n} | x_i - \bar{x} | \text{ where}$$

x_i is the individual value of the variates,

$\overline{x}$ is the arithmetic mean,

n is the number of items in the data and

$|x_i - \overline{x}|$ is an absolute value which is always taken positive.

Illustration:

Variates (x_i)	Deviation from Mean $x_i - \overline{x}$	Absolute Value $\|x_i - \overline{x}\|$
1	1 – 5 = – 4	4
3	3 – 5 = – 2	2
5	5 – 5 = 0	0
7	7 – 5 = + 2	2
9	9 – 5 = + 4	4
		$\sum \|x_i - \overline{x}\| = 12$

$\overline{x}$ = 25/5 = 5 and
Mean Deviation = 12/5 = 2.4.

In the above illustration the ignoring of + or – signs cannot be accepted in mathematical operations but in the case of mean deviation it cannot be helped as well. If the signs are not ignored, the sum of the deviations from the mean would always be zero e.g. in the preceding illustration, the total deviation from the mean would be – 4 – 2 0 + 2 and + 4 which when added together will be zero

Set I		Set II	
Variates (x_i)	$\|x_i - \overline{x}\|$	Variates (x_i)	$\|x_i - \overline{x}\|$
1	1 – 5 = 4	3	3 – 5 = 2
3	3 – 5 = 2	4	4 – 5 = 1

5	$5-5=0$	5	$5-5=0$
7	$7-5=2$	6	$6-5=1$
9	$9-5=4$	7	$7-5=2$
$\Sigma x_i = 25$	$\Sigma\lvert x_i-\bar{x}\rvert = 12$	$\Sigma x_i = 25$	$\Sigma\lvert x_i - x\rvert = 6$

Mean Deviation = 12/5 = 2.4 in Set I.

Mean Deviation = 6/5 = 1.2 in Set II.

Mean deviation provides a measure which shows by what extent the individual variates in the data deviate from the mean. In illustrations Set I and Set II, the mean in both the cases is 5 but the mean deviation in Set I is 2.4 whereas in Set II it is 1.2. In Set II where the mean deviation is less i.e. 1.2, the deviation in the individual values from the mean is less i.e. the data is more compact. On the other hand in illustration Set I where the mean deviation is more i.e. 2.4, the data is less compact. Mean is a more representative value in Set II than it is in Set I.

When the data is in grouped form then also the mean deviation can be calculated.

Illustration:

Groups	Frequency (f)	Mid–point value (x_j)	$f \times x_j$	Deviation of mid-point value from the mean $\lvert x_j-\bar{x}\rvert = \lvert d\rvert$	$f \times \lvert d\rvert$
0 – 10	4	5	20	$5-24=19$	76
10 – 20	4	15	60	$15-24=9$	36
20 – 30	6	25	150	$25-24=1$	6
30 – 40	3	35	105	$35-24=11$	33
40 – 50	2	45	90	$45-24=21$	42
50 – 60	1	55	55	$55-24=31$	31
	$\Sigma f = 20$		$\Sigma f x_j = 480$		$\Sigma f\lvert d\rvert = 224$

Mean = $\sum f x_i / \sum f = 480/20 = 24.$

Mean Deviation = $\sum f|d| / \sum f = 224/20 = 11.2.$

RELATIVE VARIABILITY

Besides Quartile Variability, Relative Varability is another form of variability which is calculated with respect to the mean deviation. It is the mean deviation as percentage to the arithmetic mean or

V_1 (Relative Variability) = Mean Deviation × 100/ Mean.

Illustration:

Variates	Mean	|Deviation|
5	12	5 – 12 = 7
7	12	7 – 12 = 5
8	12	8 – 12 = 4
11	12	11 – 12 = 1
13	12	13 – 12 = 1
18	12	18 – 12 = 6
22	12	22 – 12 = 10
Total 84		Total Deviation = 34

Mean = 84/ 7 = 12 Mean Deviation = 34/ 7 = 4.85

V_r (Relative Variability) = 4.85 x 100/12 = 485/12 = 40.4

STANDARD DEVIATION

Amongst all the measures of dispersions, the Standard Deviation is the most powerful measure of assesment. In the mean deviation, the deviations are taken irrespective of negative or positive signs which is not sound from the mathematical point of view because the signs are integral part of the values. If it has to be removed, some standradization in technique has to be adopted which can be done only by squaring the values and then adding them. This total when divided by the number of items in the data gives variance. The square root of variance is called the Standard Deviation. Hence Standard Deviation is expressed as

$$\sigma = \sqrt{\frac{\sum_{i=1}^{n} (x_i - \bar{x})^2}{n}} \quad \text{or} \quad \sqrt{\sum d^2/n}$$

Standard Deviation is therefore defined as the square root of the sum of the squares of the deviations when divided by the number of items in the data.

When there is no deviation, there is no standard deviation i.e. the variability is zero or there is no variability.

	Set A	Set B
	1	3
	2	4
	4	4
	9	5
Total	16	16
Mean =	16/4 = 4	16/4 = 4

Set A			Set B		
x_i	$(x_i - \bar{x}) = d$	d^2	x_i	$(x_i - \bar{x}) = d$	d^2
1	1 – 4 = – 3	9	3	3 – 4 = – 1	1
2	2 – 4 = – 2	4	4	4 – 4 = 0	0
4	4 – 4 = 0	0	4	4 – 4 = 0	0
9	9 – 4 = + 5	25	5	5 – 4 = + 1	1
16		38	16		2

$\bar{x} = 16/4 = 4$ $\bar{x} = 16/4 = 4$

$\sigma = \sqrt{\sum d^2/n} = \sqrt{38/4}$ $\qquad$ $\sigma = \sqrt{\sum d^2/n} = \sqrt{2/4}$

$= \sqrt{9.5} = 3.08$ $\qquad$ $= \sqrt{.5} = .7$

The set B has lesser variation in the data as compared to Set A because the standard deviation in Set B is much less than that in Set A.

The standard deviation can be calculated even when the data is in grouped form as shown below.

Illustration:

Groups	f	Mid-point value (x_j)	$f \times x_j$	$(x_j - \bar{x})$	$(x_j - \bar{x})^2$	$f(x_j - \bar{x})^2$
0 – 10	4	5	20	–16	256	1024
10 – 20	7	15	105	– 6	36	252
20 – 30	4	25	100	+ 4	16	64
30 – 40	3	35	105	+14	196	588
40 – 50	2	45	90	+24	576	1152
$\sum f = 20$			$\sum fx_j = 420$		$\sum (x_j - \bar{x})^2 = 1080$	$\sum f(x_j - \bar{x})^2 = 3080$

$\bar{x} = 420/20 = 21$

$\sigma = \sqrt{3080/20} = \sqrt{154} = 12.4$

Standard deviation can be found out by a simpler mathematical calculation in which there is no necessity of finding out the mean and the same formula for calculating the standard deviation $\sqrt{\frac{\sum (x_i - \bar{x})^2}{n}}$ can be simplified in the form of $\sigma = \sqrt{\frac{\sum x_i^2}{n} - \frac{(\sum x_i)^2}{n^2}}$

Proof:

$$\text{We have } \sigma = \sqrt{\frac{\sum (x_i - \bar{x})^2}{n}}$$

$$\sigma = \sqrt{\frac{\sum (x_i^2 - 2x_i \bar{x} + \bar{x}^2)}{n}}$$

$$\sigma = \sqrt{\frac{\sum x_i^2 - 2\bar{x}\sum x_i + \sum \bar{x}^2}{n}}$$

Since $\bar{x} = \sum x_i/n$ or $\sum x_i = n\,\bar{x}$ and

$$\sum_{i=1}^{n} \bar{x}^2 = \bar{x}^2 + \bar{x}^2 + \bar{x}^2 + \text{....n times}$$

$$= n\,\bar{x}^2$$

Substituting the value of $\sum x_i$ and $\sum \bar{x}^2$ in the above

$$\sigma = \sqrt{\frac{\sum x_i^2 - 2\,\bar{x}\,n\,\bar{x} + n\,\bar{x}^2}{n}}$$

$$= \sqrt{\frac{\sum x_i^2 - 2\,n\,\bar{x}^2 + n\,\bar{x}^2}{n}}$$

$$= \sqrt{\frac{\sum x_i^2 - n\,\bar{x}^2}{n}}$$

$$= \sqrt{\frac{1}{n}\left(\sum x_i^2 - n\,\bar{x}^2\right)}$$

$$= \sqrt{\frac{1}{n}\sum x_i^2 - \frac{n\left(\sum x_i\right)^2}{n^2}} \quad \text{since } \bar{x}^2 = \frac{\left(\sum x_i\right)^2}{n^2}$$

$$= \sqrt{\frac{\sum x_i^2}{n} - \frac{\left(\sum x_i\right)^2}{n^2}}$$

This formula provides a short cut while calculating the standard deviation.

Some of the characteristics of the standard deviation are that it makes use of the entire data; it is very suitable for algebraic calculations; its value is always definite and that is why it is used for higher statistical operations where great precision is required. It is the most powerful amongst the absolute measures of comparison.

COMBINATIONAL ANALYSIS

A modified form of standard deviation technique can be used in the classification of towns, making industrial regions, crop combination regions etc. Prof J.C. Weaver has used this technique in making crop combination regions in the Middle West.

To make a comparison of different technique of combinational analysis more understandable, the author has chosen for illustration the same table of Stone County (showing percentages under different crops) which K. Doi of Japan had chosen. Doi made certain modifications in Weaver's technique which will be dealt with later on and which gave much better results.

Weaver's Method

Illustration:

Percentage of Land Under Different Crops. Stone County

Crops	Percentages
Wheat	3.0
Barley	4.0
Corn	17.0
Oats	14.0
Oats fed unthreshed	3.0
Vegetables	5.0
Sorghum	6.0
Hay	48.0

If the area is given in acres or hectares instead of in percentages then the percentages to the total cultivated area under different crops have to be calculated and the percentages arranged in descending order as shown below:

Crops	Percentages
Hay	48
Corn	17
Oats	14
Sorghum	6

Vegetables	5
Barley	4
Wheat	3
Oats fed unthreshed	3
Total	100

If the region is a single crop region, the ideal condition would be that 100 per cent of land will be under that crop which has the maximum percentage to the cultivated area. In the above table e.g. hay has the maximum percentage of 48. It means that as far as hay is concerned, the percentage deviates by 100 – 48 = 52 per cent from the ideal condition. In case the region is double cropped, the ideal condition would be that both the crops have 50 per cent chances of occupying the total cultiaved area. Now the deviation in case of the first two leading crops from 50 per cent is to be found which is 50 – 48 = +2 in case of hay and 50 – 17 = +33 in case of corn. These deviations are then squared and added. After that the sum is divided by the number of crops in that group and the square root taken which, in a way., gives the standard deviation. The calculation of crop combination region on the basis of Weaver is illustrated below.

For mono crop

$100 - 48 = 52, \quad (52)^2 = 2704/1 \quad \sqrt{2704} = 52.0.$

For two crops

$50 - 48 = 2, \quad (2)^2 = \quad 4$

$50 - 17 = 33, \quad (33)^2 = \frac{1089}{1093/2} = 546.5$

$\sqrt{546.5} = 23.38.$

For three crops

$33.3 - 48 = -14.7, \quad (-14.7)^2 = 216.09$

$33.3 - 17 = 16.3, \quad (16.3)^2 = 265.69$

$33.3 - 14 = 19.3, \quad (19.3)^2 = 372.49$

$$\overline{854.27}/3 = 284.76$$

$$\sqrt{284.76} = 16.87$$

For four crops

$25 - 48 = -23, \quad (-23)^2 = 529$
$25 - 17 = 8, \quad (8)^2 = 64$
$25 - 14 = 11, \quad (11)^2 = 121$
$25 - 6 = 19, \quad (19)^2 = 361$

$$1075/4 = 268.75$$

$$\sqrt{268.75} = 16.39$$

For five crops

$20 - 48 = -28, \quad (-28)^2 = 784$
$20 - 17 = 3, \quad (3)^2 = 9$
$20 - 14 = 6, \quad (6)^2 = 36$
$20 - 6 = 14, \quad (14)^2 = 196$
$20 - 5 = 15, \quad (15)^2 = 225$

$$1250/5 = 250$$

$$\sqrt{250} = 15.81$$

For six crops

$16.7 - 48 = -31.3, \quad (-31.3)^2 = 979.69$
$16.7 - 17 = -0.3, \quad (-0.3)^2 = 0.09$
$16.7 - 14 = 2.7, \quad (2.7)^2 = 7.29$
$16.7 - 6 = 10.7, \quad (10.7)^2 = 114.49$
$16.7 - 5 = 11.7, \quad (11.7)^2 = 136.89$
$16.7 - 4 = 12.7, \quad (12.7)^2 = 161.29$

$$1399.74/6 = 233.29$$

$$\sqrt{233.29} = 15.27$$

For seven crops

$14.3 - 48 = -33.7, \quad (-33.7)^2 = 1135.69$
$14.3 - 17 = -2.7, \quad (-2.7)^2 = 7.29$

$$14.3 - 14 = 0.3, \quad (0.3)^2 = 0.09$$
$$14.3 - 6 = 8.3, \quad (8.3)^2 = 68.89$$
$$14.3 - 5 = 9.3, \quad (9.3)^2 = 86.49$$
$$14.3 - 4 = 10.3, \quad (10.3)^2 = 106.09$$
$$14.3 - 3 = 11.3, \quad (11.3)^2 = 127.69$$
$$1532.23/7 = 218.89$$
$$\sqrt{218.89} = 14.79.$$

For eight crops

$$12.5 - 48 = -35.5, \quad (-35.5)^2 = 1260.25$$
$$12.5 - 17 = -4.5, \quad (-4.5)^2 = 20.25$$
$$12.5 - 14 = -1.5, \quad (-1.5)^2 = 2.25$$
$$12.5 - 6 = 6.5, \quad (6.5)^2 = 42.25$$
$$12.5 - 5 = 7.5, \quad (7.5)^2 = 56.25$$
$$12.5 - 4 = 8.5, \quad (8.5)^2 = 72.25$$
$$12.5 - 3 = 9.5, \quad (9.5)^2 = 90.25$$
$$12.5 - 3 = 9.5, \quad (9.5)^2 = 90.25$$
$$1634.00/8 = 204.25$$
$$\sqrt{204.25} = 14.29$$

For all the eight crops for which the deviations have been calculated, 14.29 is the smallest figure which means that it has the least deviation. The figure that has the least deviation is supposed to be the most ideal and since this figure lies in 8 crop group, it can be concluded that the Stone County has a critical crop combination of 8 in which all the crops have to be included.

In the above calculations, the modified technique of standard deviation which is represented by $\sqrt{\sum d^2/n}$ has been used. To make the calculations still simpler, the last step may be avoided i.e. taking of the square roots. Actually the smallest square root has to be located and naturally the figure that is smallest will have the smallest square root. Therefore without taking the square root, in a step earlier, the smallest figure is obtained by dividing the sums of the squared deviation by the number of crops which gives the critical crops combination.

Doi's Method

No doubt Weaver has done pioneer work in combinational analysis which is of a very great significance, specially as far as geography is concerned, but in the first place, the process is very lengthy and secondly the results are often not too appropriate when the items are great in number. In the illustration on Weaver's technique described earlier, it can be seen that the crops like wheat and oats fed unthreshed, occupying as low as 3 per cent only have also been included in the combination region of 8 and the Stone County grows only 8 crops. This is because when we deal with the deviation in the 6, 7, 8,... crops, the sums of the squared deviations go on decreasing whereas the denominator n goes on increasing. The percentages, therefore, falling later have smallest deviations and therefore have to be included. Weaver himself has admitted this short coming in one of his articles.

Apparently from the table of Stone County it looks to be a mono-crop region or a two crop region or at the most a three crop combination region.

This discrepency has been removed upto a very great extent by K. Doi who made an improvement upon Weaver's technique. The results obtained by Doi's method are far more acceptable than those of Weaver's. Moreover Doi's method does not involve too much of calculations. Only the percentages have to be added and read on a table of critical values prepared by Doi.

According to Doi, the comparision of $\sum d^2$ for (n – 1) element combination and n-element combination is possible without calculating their actual value if $\sum_{1}^{n-1} d^2 - \sum_{1}^{n} d^2$ is found to be + or –.

Let $x_1, x_2, x_3 ....x_n$ be percentages of the first, second, third...nth ranking elements, the subtraction is expressed as follows:

$$\left(\frac{100}{n-1} - x_1\right)^2 + \left(\frac{100}{n-1} - x_2\right)^2 + \ldots . \left(\frac{100}{n-1} - x_{n-1}\right)^2 - \left(\frac{100}{n} - x_1\right)^2$$

$$- \left(\frac{100}{n} - x_2\right)^2 - \ldots . - \left(\frac{100}{n} - x_n\right)^2$$

$$= \frac{100}{n(n-1)}\left(\frac{100(2n-1)}{n} - 2\sum_{1}^{n-1} x_i\right) - \left(\frac{100}{n} - x_n\right)^2$$

Thus given n and $\sum_{1}^{n-1} x_i$ (the sum of the percentages of the 1st to (n-1) th elements) a critical value of x_n is determined which makes the zero. The table prepared by Doi lists these critical values of x_n for the necessary range of n and $\sum_{1}^{n-1} x_i$

or in the other words

$$\frac{100}{n(n-1)}\left[\frac{100(2n-1)}{n} - 2\sum_{1}^{n-1} x_i\right] - \left(\frac{100}{n} - x_n\right)^2 = 0$$

Where n is the combination, x_i is the accumulated percentage and x_n is the critical value.

Illustration:

Let us find the critical value for 6 combinations for an accumulated percentage of 70.

Substituting the values we get

$$\frac{100}{6(6-1)}\left(\frac{100(2\times 6-1)}{6} - 2\times 70\right) - \left(\frac{100}{6} - x_n\right)^2 = 0$$

$$\text{or } \frac{100}{6\times 5}\left(\frac{100(12-1)}{6} - 140\right) - (16.6667 - x_n)^2 = 0$$

$$\text{or } 3.3333\left(\frac{1100}{6} - 140\right) - (277.7789 - 33.3334x_n + x_n^2) = 0$$

$$\text{or } 3.3333(183.3333 - 140) - (277.7789 - 33.3334x_n + x_n^2) = 0$$

$$\text{or } 3.3333(43.3333) - (277.7789 - 33.3334x_n + x_n^2) = 0$$

$$\text{or } 144.4429 - 277.7789 + 33.3334x_n - x_n^2 = 0$$

$$\text{or } x_n^2 - 33.3334x_n + 133.3360 = 0$$

Now like the quadritic equation $ax^2 + bx + c = 0$ in which by Bhashkar Acharya's method $x = \frac{-b \pm \sqrt{b^2 - 4ac}}{2a}$ the above equation can be solved thus:

$$x_n = \frac{-(-33.3334) \pm \sqrt{(-33.3334)^2 - 4\times 1\times 133.3360}}{2\times 1}$$

$$= \frac{33.3334 \pm \sqrt{1111.1155 - 533.3440}}{2}$$

$$= \frac{33.3334 \pm \sqrt{577.7715}}{2}$$

$$= \frac{33.3334 \pm 24.0369}{2}$$

$$= \frac{9.2965}{2} \text{ or } \frac{57.3703}{2}$$

$$= 4.64825 \text{ or } 28.68515$$

The value of 4.64825 tallies with the table of Combinational Analysis perpared by Doi. In the same way other critical values can be calculated.

The major change that Doi has done in Weaver's technique is the replacement of $\sqrt{(\Sigma x - \bar{x})^2/n}$ by $\sqrt{(\Sigma x - \bar{x})^2}$ only. The results after this change are no doubt highly acceptable, but to avoid lengthy calculations, Doi formulated the table of combinational analysis which has been presented in the abridged form on pages 69 & 70With the help of this table only minor summations have to be done to locate the critical combination.

Illustration on Doi's Method:

Percentage of Land Under Different Crops.
Stone County

Crops	Percentages
Hay	48
Corn	17
Oats	14
Sorghum	6
Vegetables	5
Oats fed unthreshed	3
Wheat	3

Doi's Table of Combinational Analysis
(Critical Values of Combinations Against Different Percentages)

Percentages \ Combinations	2	3	4	5	6	7	8	9
50	0.00	0.00	0.00	0.00	0.00	0.00	0.00	0.00
51	1.01	0.50	0.34	0.25	0.20	0.17	0.14	0.13
52	2.04	1.02	0.68	0.51	0.40	0.34	0.29	0.25
53	3.10	1.54	1.02	0.76	0.61	0.51	0.44	0.38
54	4.17	2.06	1.37	1.03	0.82	0.68	0.59	0.51
55	5.28	2.60	1.73	1.29	1.03	0.86	0.74	0.64
56	6.41	3.15	2.09	1.56	1.25	1.04	0.89	0.78
57	7.57	3.71	2.45	1.83	1.46	1.22	1.04	0.91
58	8.77	4.27	2.83	2.11	1.69	1.40	1.20	1.05
59	10.00	4.85	3.21	2.39	1.91	1.59	1.36	1.19
60	11.27	5.44	3.59	2.68	2.14	1.78	1.52	1.33
61	12.58	6.05	3.98	2.97	2.37	1.97	1.69	1.47
62	13.94	6.67	4.38	3.27	2.60	2.16	1.85	1.62
63	15.36	7.30	4.79	3.57	2.84	2.36	2.02	1.77
64	16.83	7.95	5.21	3.88	3.09	2.56	2.19	1.92
65	18.38	8.61	5.64	4.19	3.33	2.77	2.37	2.07

66	20.00	9.30	6.07	4.51	3.59	2.98	2.54	2.22
67	21.72	10.00	6.52	4.83	3.84	3.19	2.73	2.38
68	23.54	10.73	6.97	5.17	4.11	3.41	2.91	2.54
69	25.61	11.48	7.44	5.51	4.37	3.63	3.10	2.70
70	27.64	12.25	7.92	5.86	4.65	3.85	3.29	2.87
71	30.00	13.06	8.42	6.22	4.93	4.08	3.49	3.04
72	32.68	13.90	8.93	6.58	5.22	4.32	3.69	3.22
73	35.86	14.77	9.45	6.96	5.51	4.56	3.89	3.39
74	40.00	15.69	10.00	7.35	5.81	4.81	4.10	3.58
75	50.00	16.67	10.57	7.75	6.13	5.06	4.32	3.76
76		17.70	11.16	8.17	6.45	5.33	4.54	3.95
77		18.80	11.77	8.60	6.78	5.60	4.77	4.15
78		20.00	12.42	9.05	7.12	5.87	5.00	4.35
79		21.31	13.10	9.51	7.48	6.16	5.24	4.56
80		22.79	13.82	10.00	7.85	6.46	5.49	4.78
81		24.51	14.59	10.51	8.23	6.77	5.75	5.00
82		26.67	15.43	11.06	8.64	7.10	6.02	5.23
83		30.00	16.34	11.63	9.07	7.43	5.30	5.47
84			17.36	12.25	9.52	7.79	6.60	5.72
85			18.65	12.93	10.00	8.17	6.91	5.99
86			20.00	13.68	10.52	8.57	7.24	6.27
87			22.11	14.52	11.09	9.00	7.59	6.56
88				15.53	11.72	9.48	7.97	6.88
89				16.84	12.45	10.00	8.38	7.22
90				20.00	13.33	10.60	8.84	7.60

In the first instance, all those crops are included in the combination whose accumulated percentage is less than 50, as the critical value for all elements at 50 is zero.

Now in the table the sum of the percentages under the first two crops is 48 + 17 = 65. Against this percentage when we look at the abridged table of Doi's combinational analysis, the critical value under column 3 is found to be 8.66. Since this critical value is less than the percentage of land under the third crop (Oats), it means that the region is not a 2 crop region. Therefore the pe.centage under oats is also added and it becomes 48 + 17 + 14 = 79. Now this percentage of 79, when seen against the critical value in column 4, the figure comes as 13.10, which is more than the percentage in the fourth ranking crop i.e. sorghum (6 per cent). The fourth crop, therefore, cannot be included and hence the region under study has a critical crop combination of 3 in which only hay, corn and oats are included.

If the higher ranking elements are having a very low percentages (less than 10 per cent), the lower ranking elements which have a percentage of below 5 per cent are excluded from the combination such that the occurence of fragmented combination is prevented.

Rafiullah's Method

Prof. S.M. Rafiullah's article, 'A New Approach to Functional Classification of Towns' published in The Geographer, Vol. XII, January, 1965, is a much better method than that of Weaver. But his technique also requires lengthy calculations. His method, in a way, is an improvement upon Weaver's technique. The formula evolved by Rafiullah is

$$\delta = \sqrt{\frac{\Sigma D_p^2 - \Sigma D_n^2}{N^2}}$$

where δ is the deviation, D_p the positive difference and D_n the negative difference from the medial value of the theoretical-curve value of the combination and N is the number of functions in the combination. Since it is the relative rank of the value of deviation which is needed, the square root sign may be ignored. The formula now becomes

$$\delta = \frac{\Sigma D_p^2 - \Sigma D_n^2}{N^2}$$

The maximum possible value of δ defines the critical combination.

One of the main point in this system in which it deviates from the method of Weaver is that for the first element e.g. the deviation is not taken from the ideal value which is 100 per cent, but from the middle value of the theoretical curve i.e. 50. In the same way for the second element the deviation will be from 50/ 2 = 25, in the same way for the third element it will be from 50/ 3 = 16.67 and so on.

The second point to be noted is that the deviation from the middle value of the theoretical value is in minus, but the square of the deviation will also have its prefix as minus. Thirdly the sum of the squared deviation is divided by the square of the number of elements in that group unlike its division by the actual number in that group as is done in Weaver's method. Lastly the greater is the difference, the nearer is the value to the given curve and the more critical is the combination. In Weaver's method that element is supposed to be having the critical combination in which the sum of the squared deviation when divided by the number of elements in that group is minimum.

Illustration on Rafiullah's Method:

Percentage of Land Under Different Crops. Stone County

Crops	Percentages
Hay	48
Corn	17
Oats	14
Sorghum	6
Vegetables	5
Barley	4
Oats fed unthreshed	3
Wheat	3

For mono crop

$48 - 50 = -\ 2; -(2)^2 = -4; -4/(1)^2 = -4/1 = -4$

For two crops

$$\begin{aligned}
48 - 25 &= +23; +(23)^2 = 529 \\
17 - 25 &= -\ 8; -(8)^2 = \underline{-64} \\
&= +465; +465/(2)^2 = +465/4 = +116.25
\end{aligned}$$

For three crops

$$\begin{aligned}
48 - 16.7 &= +31.3; +(31.3)^2 = +979.69 \\
17 - 16.7 &= +\ 0.3; +(0.3)^2 = +0.09 \\
14 - 16.7 &= -\ 2.7; -(2.7)^2 = \underline{-7.29} \\
&= +972.49; +972.49/(3)^2 \\
&= +972.49/9 = +108.05
\end{aligned}$$

For four crops

$$\begin{aligned}
48 - 12.5 &= +35.5; +(35.5)^2 = +1260.25 \\
17 - 12.5 &= +\ 4.5; +\ (4.5)^2 = +20.25 \\
14 - 12.5 &= +\ 1.5; +\ (1.5)^2 = +2.25 \\
6 - 12.5 &= -\ 6.5; -\ (6.5)^2 = -42.25 \\
&= +1240.50; +1240.50/(4)^2 \\
&= +1240.50/16 = +77.53
\end{aligned}$$

For five crops

$$\begin{aligned}
48 - 10 &= +38; +(38)^2 = +1444 \\
17 - 10 &= +\ 7; +\ (7)^2 = +49 \\
14 - 10 &= +\ 4; +\ (4)^2 = +16 \\
6 - 10 &= -\ 4; -\ (4)^2 = -16 \\
5 - 10 &= -\ 5; -\ (5)^2 = \underline{-25} \\
&= +1468; \quad +1468/(5)^2 = +1468/25 \\
&= +58.72
\end{aligned}$$

For six crops

$$48 - 8.3 = +39.7; +(39.7)^2 = +1576.09$$
$$17 - 8.3 = +8.7; +(8.7)^2 = +75.69$$
$$14 - 8.3 = +5.7; +(5.7)^2 = +32.49$$
$$6 - 8.3 = -2.3; -(2.3)^2 = -5.29$$
$$5 - 8.3 = -3.3; -(3.3)^2 = -10.89$$
$$4 - 8.3 = -4.3; -(4.3)^2 = -18.49$$
$$= +1649.60; +1649.60/(6)^2$$
$$= +1649.60/36 = +45.82$$

For seven crops

$$48 - 7.1 = +40.9; +(40.9)^2 = +1672.81$$
$$17 - 7.1 = +9.9; +(9.9)^2 = +98.01$$
$$14 - 7.1 = +6.9; +(6.9)^2 = +47.61$$
$$6 - 7.1 = -1.1; -(1.1)^2 = -1.21$$
$$5 - 7.1 = -2.1; -(2.1)^2 = -4.41$$
$$4 - 7.1 = -3.1; -(3.1)^2 = -9.61$$
$$3 - 7.1 = -4.1; -(4.1)^2 = -16.81$$
$$= +1786.39; +1786.39/(7)^2$$
$$= +1786.39/49 = +36.46$$

For eight crops

$$48 - 6.2 = +41.8; +(41.8)^2 = +1747.24$$
$$17 - 6.2 = +10.8; +(10.8)^2 = +116.64$$
$$14 - 6.2 = +7.8; +(7.8)^2 = +60.84$$
$$6 - 6.2 = -0.2; -(0.2)^2 = -0.04$$
$$5 - 6.2 = -1.2; -(1.2)^2 = -1.44$$
$$4 - 6.2 = -2.2; -(2.2)^2 = -4.84$$
$$3 - 6.2 = -3.2; -(3.2)^2 = -10.24$$
$$3 - 6.2 = -3.2; -(3.2)^2 = -10.24$$
$$= +1897.92; +1897.92/(8)^2$$
$$= +1897.92/64 = +29.65$$

Since the greatest difference of + 116.25 from the given curve is to be found in the two crop column so the Stone Country can be said to be having a critical crop combination of two-crops or a two-crop combination region in which the crops are hay and corn.

A comparison of the three methods in finding out the critical crop combination indicate that applying Weaver's method, Stone County has a combination of 8 crops, in Doi's technique, it comes to 3 and in Rafiullah's methods it comes to 2 only.

Experimenting these techniques in other cases it is generally found that the results obtained from Doi's technique and Rafiullah's method are very much similar but in case of Weaver there is too much of variation specially where there are sharp differences in percentages. His formula, therefore, yields widely different results. But when the percentages do not vary very much, both the formulas of Rafiullah and Weaver provide almost identical results.

CHAPTER V

PROBABILITY DISTRIBUTION

The technique and concept of probability is of an outstanding importance in statistics because it is extensively applied in the development of all physical as well as even in social sciences. It is, in a way, an index of centrality for assessing the chance of occurence as well as the effect on the chance of occurence if a change takes place in one of the conditions i.e. to determine the change. The chance of occurence can be found by observation itself but not the change due to the effect of any factor. For this purpose we require some unbiased statistical method which may determine the chance of occurence or quantum of change in one due to the change in the other which can only be a probability. The chance of the happening of an event when expressed quantitatively is called probability. It refers to a relative frequency of occurence of an event under a given set of conditions i.e. what are the chances of occurence of an event when certain conditions are given e.g. take a well-designed coin and toss it. The chances of head or tail coming up cannot be correctly predicted. There are 50 per cent chances of head coming up and 50 per cent chances of tail coming up. If it is tossed twice, the chances are that once head would be coming up and once tail coming up. But necessarily it is not so. In other words the chance is 50-50.

The probability is measured on 0 to 1 scale. Actually probability is not given a numerical value. It is never less than 0 and never more than 1.

Let there be 100 persons.

What is the probability of death in 100 persons? It is 100 per cent i.e. death is sure to occur in all the 100 persons i.e. the probability of death in a person is a must. When the probability is a surity, and it does not involve any matter of chance, the probability is said to be 1. The chances of 100 persons never dying is not possible. Therefore the probability of a person never dying is 0. If the probability is between 0 and 1, it means that it involves some level of chance and some of certainity.

The plant of wheat yielding cotton, the probability is 0 because it is impossible. The probability varies between impossible to sure i.e. from 0 to 1.

In four births, the chances are that there would be 2 boys and 2 girls i.e. the chance of 2 boys and 2 girls are 50-50 i.e. the probability is 0.5 but what would be the sequence? It may be in any of the following order :

First Birth	Second Birth	Third Birth	Fourth Birth
Girl	Boy	Girl	Boy
Boy	Girl	Boy	Girl
Girl	Girl	Boy	Boy
Boy	Boy	Girl	Girl

In all the above cases the chance is 50 per cent or the probability is 0.5. The time occurence cannot be obtained in cases of probability.

In one birth, it has to be a boy or a girl i.e. the chance is 100 per cent or probability is 1. When only two events are considered, they are binomials. There are multinomial positions also e.g. in a regular dice with 6 faces, it will have one face when cast from 1 to 6. The chance of 1 to 6 coming up when the dice is cast is 1 in 6 because it is equally likely event.

Let the faces with different numbers be as follows :

	f_1	f_2	f_3	f_4	f_5	f_6
Chances of 1 to 6 coming up	1/6	1/6	1/6	1/6	1/6	1/6

When all the faces are combined, it is 1 i.e. the probability is 1. When the dice is cast, any one face has to come up i.e. the probability of any one face coming up is 1, but which of the face would it be, cannot be predicted. Any specific face coming up will have a chance of 1/6. It is a multinomial case because there are 6 faces.

The chance of f_1 or f_2 coming up are 1/6 + 1/6 = 2/6 = 1/ 3 i.e.the probability is 1/3. It is called the Law of Addition of P (i.e probability). Similarly the chances of 1 or 2 or 3 coming up are 1/6 + 1/6 + 1/6 = 3/6 = 1/ 2 i.e. the probability is 1/ 2 = 0.5

Similarly the chances of 1 or 2 or 3 or 4 coming up are 1/6 + 1/6 + 1/6 + 1/6 = 4/6 = 2/ 3. In the same way the chances for 1 or 2 or 3 or 4 or 5 coming up are 5/6 and the probability 5/6 and similarly the chances of 1 or 2 or 3 or 4 or 5 or 6 coming up are 6/6 = 1 or a probability of 1.

The Law of Addition states that the probability of occurence of any one of the items in a selected case of 2 or more than 2 items is obtained by the algebraic sum of the probability of the individual occurence of each of the items forming the set. This law applies only to 'Either' or 'Or' situations.

Let there now be two dice of exactly similar shape each numbered from 1 to 6. When both the dice are cast the following 6 combinations may occur :

Dice 1	Dice 2	Dice 1	Dice 2	Dice 1	Dice 2	Dice 1	Dice 2	Dice 1	Dice 2	Dice 1	Dice 2
1	1	2	1	3	1	4	1	5	1	6	1
1	2	2	2	3	2	4	2	5	2	6	2
1	3	2	3	3	3	4	3	5	3	6	3
1	4	2	4	3	4	4	4	5	4	6	4
1	5	2	5	3	5	4	5	5	5	6	5
1	6	2	6	3	6	4	6	5	6	6	6

Any of the above 6 combinations are possible.

The probability of 1 coming up in both the dice is 1/36 whereas the probability of 1 coming up in only one dice was 1/6. This may be expressed as $P(f_1 \text{ of } d_1)$ is 1/6 where f_1 stands for face 1 and d_1 for dice 1. The probability of 1 coming up in dice 2 or $P(f_1\ d_2)$ is also 1/6.

$$P\ (f_1\ d_1\ f_2\ d_2) = 1/6 \times 1/6 = 1/36$$

This is the Law of Multiplication and it applies to 'And' situations e.g. one dice AND another dice. The Law of Multiplication states that the probability of the occurence of two or more than two items in a distribution taken 2 at a time is obtained by the product of the probability of individual occurences of those items in the distribution.

In the preceding table, the diagonal line XY passes through the pairs i.e. 1,1 in the first throw, 2, 2 in the second throw, 3, 3 in the third throw and so on.

When the tosses in each throws are added we get

$$1 + 1 = 2 \qquad 1 + 2 = 3$$

In the first set of second throw 2 + 1 = 3. The total is the same but the sequence is not the same. The throw-wise sum is as follows :

$d_1 + d_2$	$d_1 + d_2$	$d_1 + d_2$	$d_1 + d_2$	$d_1 + d_2$	$d_1 + d_2$
2	3	4	5	6	7
3	4	5	6	7	8
4	5	6	7	8	9
5	6	7	8	9	10
6	7	8	9	10	11
7	8	9	10	11	12

In the above table, similar totals are diagonally arranged.

When 2 dice are cast, the probability of one coming up in both i.e. a sum of 2 is 1/36. The probability of a sum of 3 is 2/36 = 1/18 because a sum of 3 can be obtained in $f_2\, d_1$ and $f_1 d_2$ combinations.

The probability for a sum of 4 is 1/36 +1/36 + 1/36 = 3/36 =1/12

The probability for a sum of 5 is 1/36 added 4 times = 4/36 = 1/9

The probability for a sum of 6 is 1/36 added 5 times = 5/36 = 1/7.2

The probability for a sum of 7 is 1/36 added 6 times = 6/36 = 1/6

The probability for a sum of 8 is 1/36 added 5 times = 5/36 = 1/7.2

The probability for a sum of 9 is 1/36 added 4 times = 4/36 = 1/9

The probability for a sum of 10 is 1/36 added 3 times = 3/36 = 1/12

The probability for a sum of 11 is 1/36 added 2 times = 2/36 = 1/18

The probability for a sum of 12 is = 1/36,

The probability of less than 2 and more than 12 is zero.

Consolidating the preceding calculations we have

The probability for a sum of 2 and 12 as 1/36,

The probability for a sum of 3 and 11 as 1/18,

The probability for a sum of 4 and 10 as 1/12,

The probability for a sum of 5 and 9 as 1/9,

The probability for a sum of 6 and 8 as 5/36,

The probability for a sum of 7 as 1/6.

In the preceding calculations, the probability distribution follows a uniform behaviour but if the behaviour is not uniform but it has even a slight deviation, then the real field of application of probability calculations is obtained. Then the actual curve is prepared which is compared with the normal frequency curve and the deviation is observed. (Fig.8).

Frequency of Rainfall in mms.

Rainfall	Frequency	Probability
160 – 180	1	1/50
180 – 200	4	4/50
·200 – 220	8	8/50
220 – 240	7	7/50
240 – 260	12	12/50
260 – 280	6	6/50
280 – 300	5	5/50
300 – 320	4	4/50
320 – 340	1	1/50
340 – 360	0	0/50
360 – 380	2	2/50

When the probability relates to a continuous number, it represents a curve but when it relates to a discrete number, it is represented by a

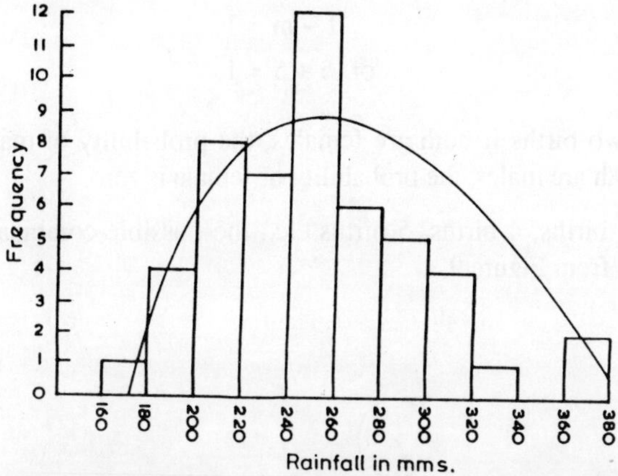

Fig.8. Histogram showing grouped frequency distribution of rainfall.

histogram (fig.8). The histogram in Figure 8 does not give a perfectly equal distribution and so the balance is not perfect. It deviates from the normal curve. It is also approaching normal curve. If the figures for a longer span (say 500 years) of rainfall are taken and plotted, it would generate a nearer to normal curve. In majority of situations when the period is increased, the curve is nearer to a perfectly normal curve. Generally they follow the normal curve law.

A distribution of probability where there can be only 2 possible outcomes, is again a very important aspect. Such a distribution is called Binomial Distribution.

In the birth of a child, the new born can be a male or a female.

In 2 births, 50 per cent chances are of a male and 50 per cent of a female.

In one birth, if the newly born is not a female, the probability of a male is 1.

In one birth, if the newly born is not a male, the probability of a female is 1.

Let the probability of a male child be m and of a female child f.

Applying the law of addition, the probability of the birth of a male or a female child is 1 i.e.

$$f + m = 1$$

$$\text{or } .5 + .5 = 1$$

In two births if both are females, the probability of male is zero and if both are males, the probability of female is zero.

In 3 births, 4 births, 5 births etc., the possible combinations can be found from Figure 9.

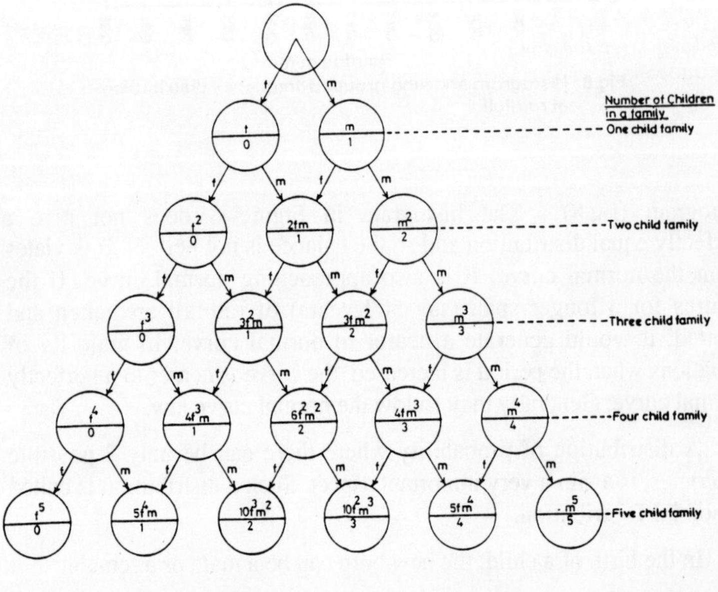

Fig.9. Probabilities associated with number of males and females in families upto five members

In the first and the last circle, the power of f and m indicates the possible combinations required. In all the circles in between, there would be m and f both. In the fourth line e.g.

$$(f^3, 3f^2m, 3fm^2, m^3)$$

indicates that the first possible combination is indicated by f^3 i.e. 3 females; the second possible combination is indicated by $3f^2m$ i.e. 2 females and 1 male; the third possible combination is indicated by $3fm^2$ i.e. 1 female and 2 males and so on. The powers in the expanded formula indicate e.g. in $3m^2f$, the power of m is 2 and of f is 1 i.e. 2 males and 1 female.

The denominator in the circle indicates the number of males.

The quotients in m^3 $3m^2f$ $3mf^2$ m^3 are 1, 3, 3 and 1 respectively. They can be very easily found out by Pascal's Triangle. The quotients actually help in calculating the probabilities.

Pascal's Triangle
(Quotients in the expansion)

For $(f+m)^0$ = 1

For $(f+m)^1$ = 1 1

For $(f+m)^2$ = 1 2 1

For $(f+m)^3$ = 1 3 3 1

For $(f+m)^4$ = 1 4 6 4 1

For $(f+m)^5$ = 1 5 10 10 5 1

For $(f+m)^6$ = 1 6 15 20 15 6 1

For $(f+m)^7$ = 1 7 21 35 35 21 7 1

For $(f+m)^8$ = 1 8 28 56 70 56 28 8 1

For the power of zero the quotient is 1,

For the power of 1, it is 1 and 1,

For the power of 2, it is 1,2 and 1,

For the power of 3, it is 1,3,3, and 1 and so on.

The Pascal's Triangle is very easy to construct. In the first and second row, l is entered. The figure in the second column of the third row is the sum of the first and the second column of the second row. The figure in the first and the last column of all the rows is 1. As a general principle, the figure in any column is the sum of the two figures above that in the preceding row e.g. in the case of $(f+m)^8$ the fifth figure of 70 is obtained by adding 35 and 35 above the row of $(f+m)^8$.

But a problem with the Pascal's Triangle is that to find the quotient of any item, the whole of the Pascal's Triangle is to be made which is a very cumbersome procedure. The quotients can, therefore be very easily obtained by the Permutation and Combination technique for which the formula used is as follows :

$$^{n}C_{r} = \frac{n\,!}{r\,!\,(n-r)\,!}$$

Where n stands for the power for which the quotient is to be found and r stands for the term required e.g.

for $(x + y)^{10}$ the first term will be $^{10}C_1 = \frac{10\,!}{1\,!\,(10-1)\,!}$

$$= 10! / 9! = 10 \times 9! / 9! = 10$$

The second term will be $^{10}C_2 = \frac{10\,!}{2\,!\,(10-2)\,!}$

$$= \frac{10 \times 9 \times 8\,!}{2 \times 8\,!}$$

$$= 45 \text{ and so on.}$$

In probability, there are two types of operations. One is the principle consistent operation e.g. the latitude of Aligarh which is fixed and not changing. Therefore the effect of latitude on rainfall is consistent. The height of Aligarh above sea level is also consistent, the distance of Aligarh from the sea from a particular point is also consistent, therefore the rainfall at Aligarh must also be consistent every year. But in all these situations, there are some erratic and random factors which affect the amount of rainfall e.g. the track of a particular disturbance, any abnormal phenomenon in sun spots or the formation of a high pressure or a low pressure area. They are random or chance factors. The magnitude of these factors can be found out by probability. In random sampling also the magnitude of randomness is also checked by probability. The level of confidence can also be found out by probability.

The other operation is how to bind discrete numbers.

Let there be a family of 6 children.

The chances of all being male are m^6.

The chances of all being female are f^6.

Expanding $(m+f)^6$ we get

$$m^6 + 6m^5f + 15m^4f^2 + 20m^3f^3 + 15m^2f^4 + 6mf^5 + f^6$$

Since there are two factors only i.e. male or female, the probability of a male can be only .5 and that of a female also .5. Instead of .5 we can also take 1/ 2. Now substituting 1/ 2 for m and f we get

$$\text{Probability} = (1/2)^6 + 6(1/2)^5 \times 1/2 + 15(1/2)^4 \times (1/2)^2 + 20(1/2)^3 \times (1/2)^3 + 15(1/2)^2 \times (1/2)^4 + 6(1/2) \times (1/2)^5 + (1/2)^6$$

$$= 1/64 + 6/64 + 15/64 + 20/64 + 15/64 + 6/64 + 1/64$$

The arithmetic mean in this perfectly balanced distribution is of 3 males and 3 females i.e. 20/64.

Number of male or female child (1)		0	1	2	3	4	5	6
Probability (2)	=	1/64	6/64	15/64	20/64	15/64	6/64	1/64
(1) × (2)	=	0	6/64	30/64	60/64	60/64	30/64	6/64

$$\text{Mean} = 0 + \frac{6}{64} + \frac{30}{64} + \frac{60}{64} + \frac{60}{64} + \frac{30}{64} + \frac{6}{64}$$

$$= \frac{0 + 6 + 30 + 60 + 60 + 30 + 6}{64}$$

$$= 192/64 = 3$$

The main features of such a distribution are

1. The probability distribution is perfectly symmetrical.
2. The distribution is bisected by the arithmetic mean i.e. equal probability occurs on either side of the arithmetic mean.
3. The cone is found highest around the arithmetic mean (Fig. 10).

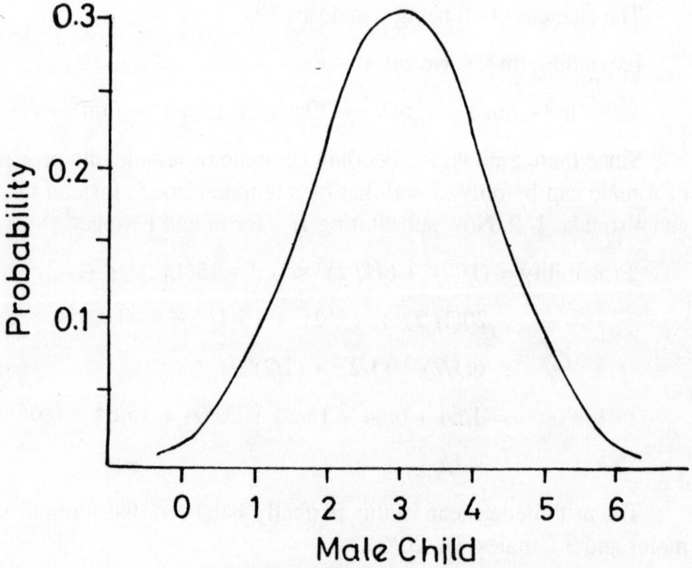

Fig.10. Probability of male child in 6 births

The mean of occurence of male in a family of 6 persons in this case is 3 but necessarily it is not so. In a survey of 10 families each with 6 children, if the total number of males are divided by 10 it would come around 3 (may be 3 also). The greater the number of families surveyed, the nearer one would approach mean.

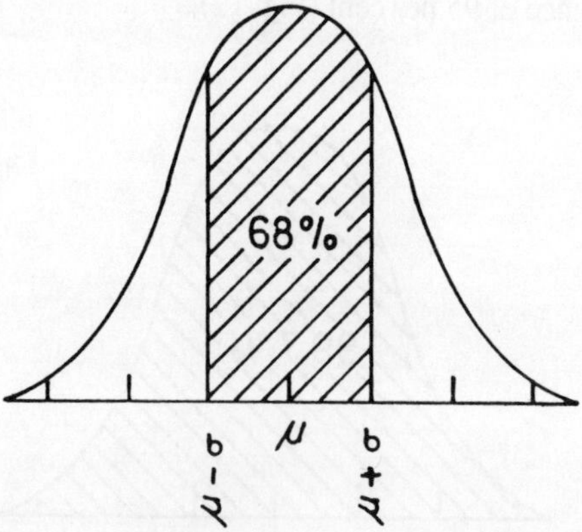

Fig.11 Area under normal curve within one standard deviation of the mean.

Let μ be the arithmetic mean and

σ the standard deviation.

The area lying between $\mu - \sigma$ and $\mu + \sigma$ and the corresponding curve i.e. between one standard deviation range has a chance occurence of 68 per cent (Fig.11).

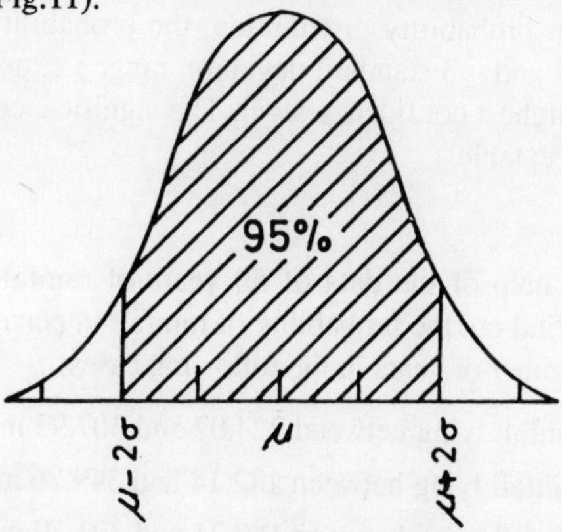

Fig.12 Area under normal curve within two standard deviation of the mean.

Similarly the area lying between μ + 2σ and μ – 2σ range has a chance occurence of 95 per cent (Fig. 12).

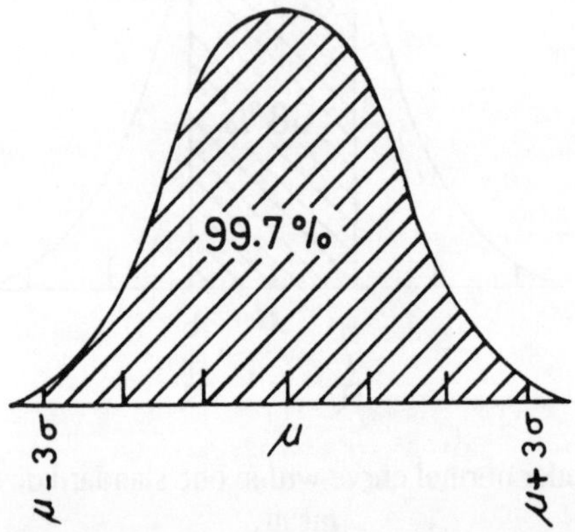

Fig. 13 Area under normal curve within three standard deviation of the mean.

In the same way the area lying between μ + 3 σ and μ – 3 σ range has a chance occurence of 99.7 per cent (Fig. 13).

Usually in probability distribution, the probability occurs within the limit of +3 and – 3 standard deviation range i.e. upto 99.7 per cent which is the highest confidence level. The significance level of 95 per cent is just acceptable.

Illustration :

With the help of the data of 50 years of rainfall in millimetres given below, find out the probability of rainfall in percentage as well as in absolute number of years in the following cases:

a) The rainfall lying between 224.07 and 307.93 mms.

b) The rainfall lying between 182.14 and 349.86 mms.

c) The rainfall lying between 140.21 and 391.79 mms.

313	380	241	267	251	252	285	202	257	346
281	268	249	296	225	319	227	259	289	298

247	268	285	311	243	255	206	223	226	303
272	234	380	252	232	188	251	286	222	251
234	258	328	286	298	245	219	284	208	300

Solution : $\sigma = 41.93$ and $\mu = 266$

a) 1 standard deviation range on the negative side =266 – 41.93 = 224.07 mms.

1 standard deviation range on the positive side = 266 + 41.93 = 307.93 mms.

i.e. 1 standard deviation range is from 224.07 to 307.93 mms. which is the same as is given in question part (a) itself. Since the value lies in 1 standard deviation range, the chance of occurence is 68 per cent or .68 (See Fig.11).

As far as absolute number of years are concerned, it would be $.68 \times 50 = 34.00$ years.

b) 2 standard deviation range on the negative side is 224.07 – 41.93 = 182.14 mms.

2 standard deviation range on the positive side is 307.93 + 41.93 = 349.86 mms.

i.e. the 2 standard deviation range in total is 182.14 to 349.86 mms. as is given in the (b) part of the question itself. Since it lies in the 2 standard deviation range, the chance of occurence is 95 per cent or .95 (See Fig.12).

As far as the absolute number of years are concerned, it would be $.95 \times 50 = 47.50$ years.

c) 3 standard deviation range on the negative side is 182.14 – 41.93 = 140.21 mms.

3 standard deviation range on the positive side is 349.86 + 41.93 = 391.79 mms.

i.e. the total standard deviation range is from 140.21 to 391.79 mms. as has been asked in the question itself. Since it lies in the 3 standard deviation range, the chance of occurence is 99.7 per cent or .997 (See Fig.13).

As far as absolute number of years are concerned, it would be $.997 \times 50 = 49.85$ years.

In the data given, the actual number of years having rainfall between 224.07 to 307.93 mms. or approximately 224 to 308, is 37 whereas according to our calculations it comes as 34 years which is approximately correct.

Similarly in part (b), it comes as 47.5 years whereas actually it is 48 years which is also approximately correct and in part (c), the value comes as 49.85 whereas actually it is 50 years. All the three results very nearly tally with the actual situation.

When 1, 2, and 3 standard deviation ranges are defined it is quite easy to calculate the probability but if the figures do not come exactly in the standard deviation ranges for which normal probability curves are used, some other methods have to be adopted. The values of probability seldom follow the normal curve and the distribution is skewed and highly generalised–when an event is likely to occur or that when that event is not likely to occur e.g. we want to know the number of rainy days in the months of rainy season. We are not interested to know the number of days when the rain does not fall. We are actually interested in knowing the number of days on which the rain falls during the rainy season of a month.

In this connection one is faced with two problems :

a) The first problem is to determine the probability of an event occuring with a value lying between an arithmetic mean and the standard deviation. For this purpose one has to calculate the z-score (zee score) and then trace the value of probability of the z-score from the normal curve.

$$\text{z - score} = \frac{x - \mu}{\sigma}$$

where x is the value for which the probability of occurence is to be determined,

μ is the arithmetic mean and

σ is the standard deviation.

Illustration :

What is the probability of a rainfall of not more than 270 mms. or less than 270 mms. when the mean is 265 and the standard deviation 4.5?

$$z\text{-score} = \frac{x-\mu}{\sigma} = \frac{270-265}{4.5} = \frac{5}{4.5} = 1.11$$

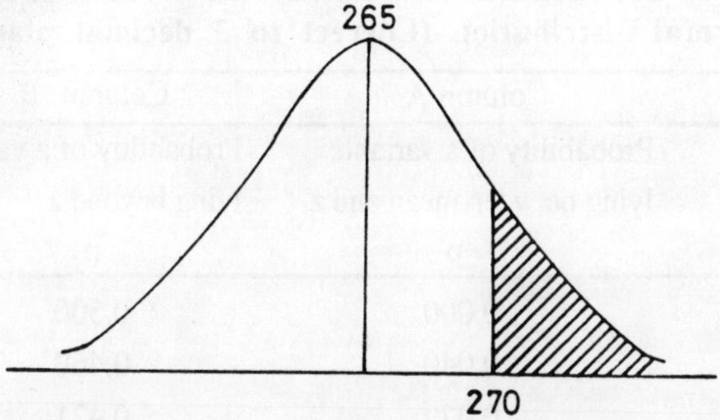

Against this z-score of 1.11, the figure of probability in p column of table of probabilities associated with values of z is 0.364. Therefore it is concluded that the probability of rainfall of not more than 270 mms. is 36.4 per cent

If suppose the probability of rainfall of a range say more than 50 mms. and less than 75 mms. is to be found out when the standard deviation is 4.5 and the mean is 60, first of all the probability of more than 50 mms. of rainfall and then the probability of less than 75 mms. of rainfall has to be found out and then added together e.g. for a rainfall of more than 50 mms.

$$z\text{-score} = \frac{x-\mu}{\sigma} = \frac{50-60}{4.5} = \frac{-10}{4.5} = -1.11$$

Similarly

z-score for less than 75 mms. of rainfall would be

$$\frac{75-60}{4.5} = \frac{15}{4.5} = 3.33.$$

The probability of 1.11 z-score is 0.364 and

the probability for 3.33 z-score is 0.499.

Adding them together we get the probability for the whole range of rainfall between not more than 50 mms. and not less than 75 mms. as 0.863 i.e. 86.3 per cent.

Table of Probabilities Associated with Values of z in a Normal Distribution (Correct to 3 decimal places)

	Column A	Column B
Z-score	Probability of a variable lying between mean and z p	Probability of a variable lying beyond z p_1
0.0	0.000	0.500
0.1	0.040	0.460
0.2	0.079	0.421
0.3	0.118	0.382
0.4	0.155	0.345
0.5	0.191	0.309
0.6	0.226	0.274
0.7	0.258	0.242
0.8	0.288	0.212
0.9	0.316	0.184
1.0	0.341	0.159
1.1	0.364	0.136
1.2	0.385	0.115
1.3	0.403	0.097
1.4	0.419	0.081
1.5	0.433	0.067
1.6	0.445	0.055
1.7	0.455	0.045
1.8	0.464	0.036
1.9	0.471	0.029
1.96	0.475	0.025
2.0	0.477	0.023

2.1	0.482	0.018
2.2	0.486	0.014
2.3	0.489	0.011
2.4	0.492	0.008
2.5	0.494	0.006
2.58	0.495	0.005
2.6	0.495	0.005
2.7	0.496	0.004
2.8	0.497	0.003
2.9	0.498	0.002
3.0	0.499	0.001
3.1	0.499	0.001
3.2	0.499	0.001
3.3	0.499	0.001
3.4	0.500	0.000
3.5	0.500	0.000

The second problem is of solving the querry as to what is the probability of occurence of an isolated event in a distribution which is highly skewed? It can be solved only with the help of Poisson's Distribution. It is based upon a series of binomial expansion of $\left(1+\frac{1}{n}\right)^n$. It is a very standard technique of binomial algebra.

$$\left(1+\frac{1}{2}\right)^2 = 1 + 1 \times \frac{1}{2} \times 2 + \frac{1}{4} = 1 + 1 + \frac{1}{4} = \frac{4+4+1}{4} = 9/4 = 2.25$$

$$\left(1+\frac{1}{3}\right)^3 = 1 + 3 \times 1^2 \times \frac{1}{3} + 3 \times 1^1 \times \left(\frac{1}{3}\right)^2 + \left(\frac{1}{3}\right)^3 = 1 + 1 + \frac{3}{9} + \frac{1}{27}$$

$$= \frac{27+27+9+1}{27} = \frac{64}{27} = 2.37$$

$$\left(1+\frac{1}{4}\right)^4 = 1 + 4 \times 1^3 \times \frac{1}{4} + 6 \times 1^2 \times \left(\frac{1}{4}\right)^2 + 4 \times 1^1 \times \left(\frac{1}{4}\right)^3 + \left(\frac{1}{4}\right)^4$$

$$= 1 + 1 + \frac{6}{16} + \frac{4}{64} + \frac{1}{256} = \frac{256+256+96+16+1}{256}$$

$$= \frac{625}{256} = 2.44$$

Now if n is increased upto infinity, it will be 2.7183 and it is represented by 'e' i.e. e = 2.7183 which is very extensively used in connection with Poisson's Distribution. In e^{-z}, z stands for mean in a skewed distribution.

Exercise: With the help of the table given below, calculate the probability of 0, 1, 2, 3, 4, 5, and 6 droughts per year.

Number of Years	Number of Droughts Per Year	Total Droughts
18	0	0
25	1	25
11	2	22
12	3	36
19	4	76
9	5	45
6	6	36
100		240

Solution:

The above table indicates 240 droughts in 100 years i.e. 2.4 droughts per year which is represented by z. Also certain years did not have any drought i.e. they were zero years as far as droughts were concerned.

To calculate the probability of the number of droughts, the formula will be as follows:

$$\text{Probability of number of droughts} = e^{-z} \times \frac{z^x}{x!} \quad \text{where}$$

$$e^{-z} = \frac{1}{e^z} = \frac{1}{(2.7183)^{2.4}} \text{ (since there are 2.4 droughts per year)}$$

= 0.0907165

z is number of droughts per year,

x is number of droughts for which calculation is required,

e is a constant the value of which is 2.7183.

Droughts (x)	e^{-z} (1)	$\frac{z^x}{x!}$ (2)	Probability (1)×(2)	Percentage
0	0.0907165	$\frac{(2.4)^0}{0!}$	0.0907165	9.07165
1	0.0907165	$\frac{(2.4)^1}{1!}$	0.2177196	21.77196
2	0.0907165	$\frac{(2.4)^2}{2!}$	0.2612635	26.12635
3	0.0907165	$\frac{(2.4)^3}{3!}$	0.2090108	20.90108
4	0.0907165	$\frac{(2.4)^4}{4!}$	0.1254065	12.54065
5	0.0907165	$\frac{(2.4)^5}{5!}$	0.0601951	6.01951
6	0.0907165	$\frac{(2.4)^6}{6!}$	0.0240781	2.40781
		Total	0.9883901	98.83901

The calculations approximately tally with the actual figures in the table. In such a distribution which is highly skewed only the frequency of droughts can be calculated. It cannot be predicted as to when the droughts are likely to occur. This is entirely a separate problem.

CHAPTER VI

SAMPLING

Sampling briefly means the study of a small or selected proportion of the item of population. But in such a selection certain restraint has to be applied e.g. the small selection should be such that it is reasonably representative of the total population or parent population or the universe. In such a selection two things must be clear. First e.g. if 10 is the sample drawn, it is composed of the same subject. Secondly that if out of 200 tables, 100 are brown, 40 are black and 60 green, the subject here has a distinctive character e.g. some are brown, some are black and some are green. In the sample of 10 the number of tables in each colour is not known. Suppose 50 per cent tables are brown, 20 per cent black and 30 per cent green, it is not necessary that the sample of 10 drawn may have the same percentages in respect of colour. Suppose the sample of 10 drawn has 4 black tables, 3 brown and 3 green, it means that the sample enjoys all the characteristics as far as colours are concerned.

Parameter refers to the characteristics of the total population whereas statistics refers to the characteristics of sample e.g. arithmetic mean, variance, standard deviation etc. of the total population are parameters and that of the sample are statistics. The arithmetic mean in statistics is represented by $\overline{x}$ while in parameter it is represented by μ (mew). Variance, standard deviation and correlation in parameter are represented as V,σ (sigma) and ρ (rho) respectively whereas in statistics they are represented by s^2, s and r.

The need in sampling may be either imposed or may be by choice. Suppose the rainfall distribution is to be studied and the average rainfall of a particular station is to be known. This average could be from a record of 30 years, 50 years or even 100 years. The rainfall has not been recorded since it has been falling. Therefore we are under compulsion to take samples for 30, 50 or 100 years i.e. the need in this case has been imposed upon us because the rainfall records since when it has been falling (say 1,00,000 years) cannot be made available.

The need may be choice based. Let there be 300 villages in 3 districts to be surveyed but the time is limited. The 300 villages when surveyed will take a very long time. Suppose the time is unlimited then too it would be a very laborious job to survey all the 300 villages. Hence for the sake of convenience, sample would be drawn such that they are truely representative of the entire data.

Arithmetic mean is obtained by adding say 30 years of annual rainfall and dividing it by 30. From this statement it is inferred that the distribution of total rainfall is such that it has the same average rainfall for 30 years. To state that the rainfall of a certain station is 25 cms. based on a record of 30 years would be a very correct statement. But to say that the average is 25 cms. is not very correct statistically because in this case the period for which the rainfall has been recorded has not been taken into consideration. Sample is always used as an instrument to infer within certain limits. The inference is made from the sample characteristic or total population characteristic. The entire population can neither be generated nor used. Sample is a reasonable representation of total population if properly drawn.

Methods of Sampling

The standard deviation indicates the nature of the distribution of the variates. Deviation from the normal curve indicates skewness when plotted on any scale.

The following data when plotted gives a normal curve (Fig. 14):

Years	Rainfall (mms.)
1900	1
1901	2
1902	3
1903	4
1904	4
1905	3
1906	2
1907	1

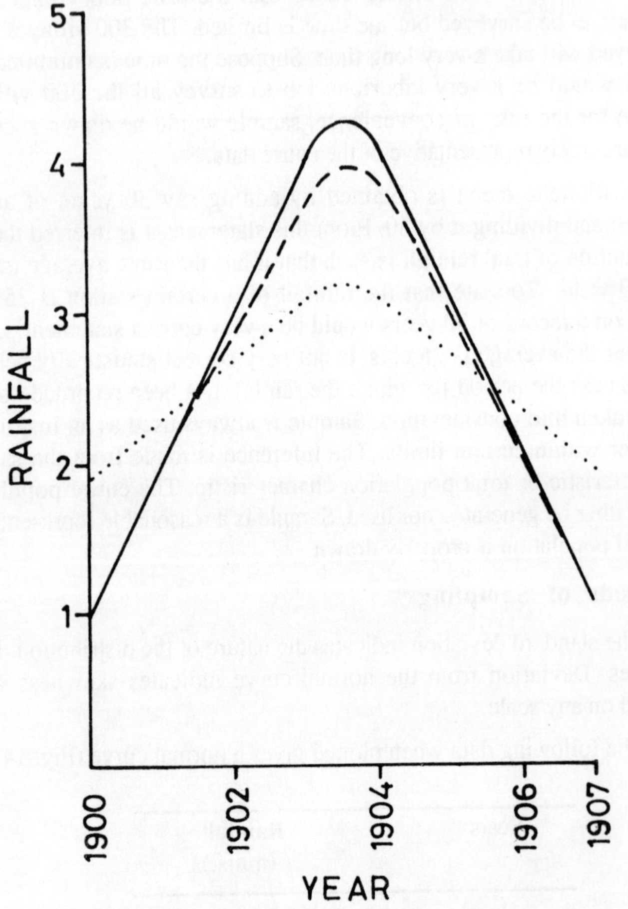

Fig.14. Distribution curve showing relationship between sample size and frequency curve

Instead of this if the mean values of first second, third fourth, fifth sixth and seventh eight i.e. 1.5, 3.5, 3.5 and 1.5 respectively are plotted on the same graph and on the same scale, the shape of the line graph would be as is indicated by broken line and the same line graph will be narrower (Fig.14). If only two values resulting from the mean of the first four values i.e. 2.5 and the mean of the last four values i.e. again 2.5 is plotted on the same graph and on the same scale as is shown by dotted line (Fig.14), the line graph becomes still narrower. It means that the narrowness of the graph indicates nearness to the mean implying that the representative characteristic of the sample increases directly with the size of the sample i.e. the larger the sample, the more it is representative of the total population and vice versa. According to the Law of Large Numbers the tendency of a random sample distribution, to resemble its parent population, increases with the increase in the size of the sample. The statistics reaches the parameter as the size of the sample increases. It is expressed as s ⟶p with the increase in the size of the sample.

$\sigma = \sqrt{\frac{\Sigma d^2}{n}}$ is not for the total population because the total population is not known and cannot be known. When we mention 50 years of data, it is also assumed that the rainfall has been falling for more than 50 years although the data is only for 50 years. We want to have the best possible of the total population. But we have to infer from the data as to what would be the total population. The estimate should be unbiased which can be found by Bessel's correction for variance and which is equal to n/(n -1). When the standard deviation $= \sqrt{\frac{\Sigma d^2}{n}}$, variance is $= \frac{\Sigma d^2}{n}$. Applying Bessel's correction we have

$$\text{Variance} = \frac{\Sigma d^2}{n} \times \text{Bessel s correction}$$

$$= \frac{\Sigma d^2}{n} \times \frac{n}{(n-1)}$$

$$= \Sigma d^2 / (n-1)$$

Illustration:

Let Variance = 16 for 50 years; S.D. = 4. $\Sigma d^2 = 800$

Applying Bessel's correction we have

$$\text{V ariance} = \frac{800}{(50-1)}$$

$$= \frac{800}{49}$$

or the best estimate standard devitation

$$= \sqrt{\frac{16 \times 50}{49}}$$

$$= \frac{4}{7}\sqrt{50}$$

$$= 4.04$$

In all sample population to make the result unbiased, the denominator should always be (n – 1) and not n i.e. $\sqrt{\Sigma d^2/(n-1)}$ but frequently this is not done.

There are two broad types of sampling:

1 Probability sampling done randomly and

2 Purposive or judicious sampling which is of non-random type.

1. Probability Sampling

Probability sampling is the most frequently used category of sampling under which four sub-types have been recognized:

a) Unrestricted Random Sampling

b) Restricted Random Sampling

c) Serial Random Sampling and

d) Proportional Probability Sampling

In each of the above four sub-types, the word random is common. The classified defination of random is that it is that process of sample drawing in which each item of the population stands equal chance of selection. As the time passed, new types of data emerged and new systems of statictics developed and it was found that the above definition of random did not operate fully and did not convey the sense of the term. Hence at an International Conference in Statistics held in 1957, it was suggested that the definition be modified and hence it was adopted that the random process should be defined as follows:

A process can be described as random if each item of the population is assigned, initially, a determined known probability of it being selected. The idea of randomness was the equality of chance of selection in which the probability technique is used.

a) Unrestricted Random Sampling

As its name indicates, no restrictions are applied in it. The best example is that if a coin is tossed, 50 per cent chances are that tail would turn up. In a dice with 6 faces, every face has a chance of turning up 1/6 times which is only a theoritical expectation. The figure of 1 may appear twice or thrice in 6 throws. In 6 throws it is not necessary that all the 6 numbers from 1 to 6 will appear one by one and in 600,000 throws, 100,000 times each face will turn up. As the number of throws increases, the probability approaches 1/6.

The chances of ith item of selection are equal to the chances of jth item of selection i.e. if there are 5 items in a group

$$P_{i(5)} = P_{j(5)}$$

The above equation in the generalised form is expressed as

$P_{ik} = P_{jk}$ which means that ith group consisting of k items has equal chances of selection as the jth group consisting of k items.

Here there is no restriction.

b) Restricted Random Sampling

This type of sampling is very useful in geography specially stratified random sampling.

Let there be three groups in population of n items (n > 64).

Group I	0 – 20 items
Group II	21 – 64 items
Group III	65 – n items

A sample of equal numbers (say per cent) from each strata is to be drawn. But 10 per cent sample in each strata will not be the same as far as the number of samples drawn is concerned out of the total population. Therefore the absolute size of the sample drawn will not be equal.

Suppose 1000 samples are to be drawn from each strata. The number of samples drawn are the same but they are not proportional to the total population. If the size of the sample is equal with respect to total population, it will not be equal to the sample in each strata. For this restriction, the method is called 'Restricted' random sampling.

The strata of equal size can be generated e.g. making crop pattern from 100 villages.

Let there be four different soil types, the area under each soil zone being almost equal, and let the soil zones be S_1, S_2, S_3, S_4 with each zone having the same number of villages.

The total size of sample is say 20 villages i.e. from each soil zones, 5 samples are to be drawn i.e. in all, 20 samples from 100 villages. This 20 out of a total of 100 villages will be quite a good representation with a high degree of confidence. Any findings obtained on the basis of these samples will be considered quite legitimate.

This type of sampling is a sort of group sampling in which all items do not have equal chances of selection if the samples are drawn without making strata.

Let the total population be of size N and a random sample of size n is to be selected from it.

There are

$$\frac{N!}{n!(N-n)!} \quad \text{or } {}^{N}C_{n}$$

i.e. a combination of N taken n at a time possible sample.

The probability of selection is

$$\frac{1}{{}^{N}C_{n}} = \frac{1}{\frac{N!}{n!(N-n)!}}$$

Let N be 100 and n = 5

The probability is

$$= \frac{1}{\frac{100!}{5!(100-5)!}}$$

$$= \frac{1}{100!} \times 5!(100-5)!$$

$$= \frac{5! \times 95!}{100!}$$

$$= \frac{5! \times 95 \times 94 \times 93........\times 1}{100 \times 99 \times 98 \times 97 \times 96 \times 95 \times 94 \times ...... \times 1}$$

$$= \frac{5 \times 4 \times 3 \times 2 \times 1}{100 \times 99 \times 98 \times 97 \times 96}$$

$$= \frac{1}{75{,}287{,}520}$$

$$= 0.0000000125 \text{ or}$$

in terms of percentage it would be 0.00000125 per cent. i.e. the probability of selecting n out of N units is the same as 0.00000125 per cent

There is no hard and fast rule regarding the percentage of the size of sample.

c) Serial Random Sampling

Let there be 100 plots numbered 1 2 3 4 5 100

One has to decide from the point from which plot the sampling has to be started and what the interval has to be chosen.

Let us start from the Plot No. 4 and let the interval be 5.

The samples thus would be 4 9 14 19 and so on.

It means that only the interval has to be selected and the randomness is only in selecting the starting point. The method being not very sophisticated is generally not used if delicate problems are to be taken.

d) Proportional Probability Sampling

Class	Elements	Value of Elements	Total
A	4	3 4 5 6	18
B	2	1 2	3
	6	6	21

In A there are 4 elements 3 4 5 6

In B there are 2 elements 1 2

Total of elements in A = 18

Mean of elements in A = 18/4 = 4.5

Total of elements in B = 3

Mean of elements in B = 3/ 2 = 1.5

The number of elements in A and B are unequal.

A has greater number of elements i.e. it has greater proportion.

The chance of probability in A must be 4/6 = 2/ 3.

The chance of probability in B must be 2/6 = 1/ 3.

2/ 3 + 1/ 3 = 1

The probability can never be more than 1.

In A the mean value in the given probability = 2/ 3 × 4.5 = 3

In B the mean value in the given probability = 1/ 3 × 1.5 = 0.5

Total of A and B = 3 + 0.5 = 3.5 which is the total probability with respect to mean value.

3.5 × 6 (Number of elements) = 21 (Total of the values of elements).

i.e. Total number of elements =

Number of elements × Total Probability with respect to the mean value.

This type of proportional probability sampling is not used in geographical studies usually.

2. Purposive Sampling

Purposive sampling is done in accordance with some specific purpose behind it.

Let P1, P2, P3.........be the population of the villages to be surveyed. A survey is to be carried on in the villages which have a population of more than 500 persons. Suppose S_3 or any other village has a population below 500, it will not be taken into consideration because it has no significance as far as our problem is concerned. Since our purpose behind the problem is not fulfilled, therefore S_3 village is discarded.

How to draw samples:

To draw the sample there are three broad categories of problem:

1) The problem which concerns with the individual item in a distribution or operation;

2) The category of problems which deal with some spatial location in a population and

3) The problem which deals with unit areas of a population.

Suppose unrestricted sampling is to be carried on in 100 villages and 20 samples are to be drawn. Assign a number to each of the village like 00 01 02 03 04 ...... 99. With the help of the table of Random Sampling Numbers (pages 105 to 107), start from anywhere and choose the numbers as shown in the table in the same order so that 20 numbers are selected. The numbers that are being repeated are discarded. The numbers so drawn do not have any subjective element in their selection and they are without any bias or the selection is unrestricted. Depending upon the starting point in drawing the samples, different persons will start from different points or numbers so that the samples drawn will differ from person to person. In this case let us start from 34. Counting 20 samples starting from 34 (row 16, column 6), we will have

34 55 26 08 39 73 27 40 33 79 36 59 88 80 20
55 41 13 77 07 40 30

Random Sampling Numbers

20 17 42 28 23 17 59 66 38 61 02 10 86 10 51 55 92 52
74 49 04 49 03 04 10 33 53 70 11 54 48 63 94 60 94 49
94 70 49 31 38 67 23 42 29 65 40 88 78 71 37 18 48 64
22 15 78 15 69 84 32 52 32 54 15 12 54 02 01 37 38 37
93 29 12 18 27 30 30 55 91 87 50 57 58 51 49 36 12 53

45 04 77 97 36 14 99 45 52 95 69 85 03 83 51 87 85 56
44 91 99 49 89 39 94 60 48 49 06 77 64 72 59 26 08 51

16 23 91 02 19 96 47 59 89 65 27 84 30 92 63 37 26 24
04 50 65 04 65 65 82 42 70 51 55 04 61 47 88 83 99 34
32 70 17 72 03 61 66 26 24 71 22 77 88 33 17 78 08 92

03 64 59 07 42 95 81 39 06 41 20 81 92 34 51 90 39 08
62 49 00 90 67 86 93 48 31 83 19 07 67 68 49 03 27 47
61 00 95 86 98 36 14 03 48 88 51 07 33 40 06 86 33 76
89 03 90 49 28 74 21 04 09 96 60 45 22 03 52 80 01 79
01 72 33 85 52 40 60 07 06 71 89 27 14 29 55 24 85 79

27 56 49 79 34 34 32 22 60 53 91 17 33 26 44 70 93 14
49 05 74 48 10 55 35 25 24 28 20 22 35 66 66 34 26 35
49 74 37 25 97 26 33 94 42 23 01 28 59 58 92 69 03 66
20 26 22 43 88 08 19 85 08 12 47 65 65 63 56 07 97 85
48 87 77 96 43 39 76 93 08 79 22 18 54 55 93 75 97 26

08 72 87 46 75 73 00 11 27 07 05 20 30 85 22 21 04 67
95 97 98 62 17 27 31 42 64 71 46 22 32 75 19 32 20 99
37 99 57 31 70 40 46 55 46 12 24 32 36 74 69 20 72 10
05 79 58 37 85 33 75 18 88 71 23 44 54 28 00 48 96 23
55 85 63 42 00 79 91 22 29 01 41 39 51 40 36 65 26 11

67 28 96 25 68 36 24 72 03 85 49 24 05 69 64 86 08 19
85 86 94 78 32 59 51 82 86 43 73 84 45 60 89 57 06 87
40 10 60 09 05 88 78 44 63 13 58 25 37 11 18 47 75 62
94 55 89 48 90 80 77 80 26 89 87 44 23 74 66 20 20 19
11 63 77 77 23 **20** 33 62 62 19 29 03 94 15 56 37 14 09

64 00 26 04 54 55 38 57 94 62 68 40 26 04 24 25 03 61
50 94 13 23 78 41 60 58 10 60 88 46 30 21 45 98 70 96
66 98 37 96 44 13 45 05 34 59 75 85 48 97 27 19 17 85
66 91 42 83 60 77 90 91 60 90 79 62 57 66 72 28 08 70
33 58 12 18 **02** 07 19 40 21 29 39 45 90 42 58 84 85 43

52 49 40 16 72 40 73 05 50 90 02 04 98 24 05 30 27 25
74 98 93 99 78 30 79 47 96 **92** 45 58 40 37 89 76 84 41
50 26 54 30 01 88 69 57 54 45 69 88 23 21 05 69 93 44
49 46 61 89 33 79 96 84 28 34 19 35 28 73 39 59 56 34
19 65 13 44 78 39 73 88 62 03 36 00 25 96 86 76 67 90

64 17 47 67 87 59 81 40 72 61 14 00 28 28 55 86 23 38
18 43 97 37 68 97 56 56 57 95 01 88 11 89 48 07 42 60
65 58 60 87 51 09 96 61 15 53 66 81 66 88 44 75 37 01

79 90 31 00 91 14 85 65 31 75 43 15 45 93 64 78 34 53
07 23 00 15 59 05 16 09 94 42 20 40 63 76 65 67 34 11

90 08 14 24 01 51 95 46 30 32 33 19 00 14 19 28 40 51
53 82 62 02 21 82 34 13 41 03 12 85 65 30 00 97 56 30
98 17 26 15 04 50 76 25 20 33 54 84 39 31 23 33 59 64
08 91 12 44 82 40 30 62 45 50 64 54 65 17 89 25 59 44
37 21 46 77 84 87 67 39 85 54 97 37 33 41 11 74 90 50

The numbers 55 and 40 have been selected once only because the numbers that have appeared more than once are discarded.

The table of random sampling numbers which may be of two digits column may be used for drawing samples for three, four, five etc. digits by combining more columns. In the above case since the total number of villages are 100 i.e. 00 to 99, the table with even two digits has sufficed.

If the samples are drawn to show locational factors e.g. areas etc., then a mesh of horizontal and vertical lines is drawn in such a way that each line is assigned a number as shown below:

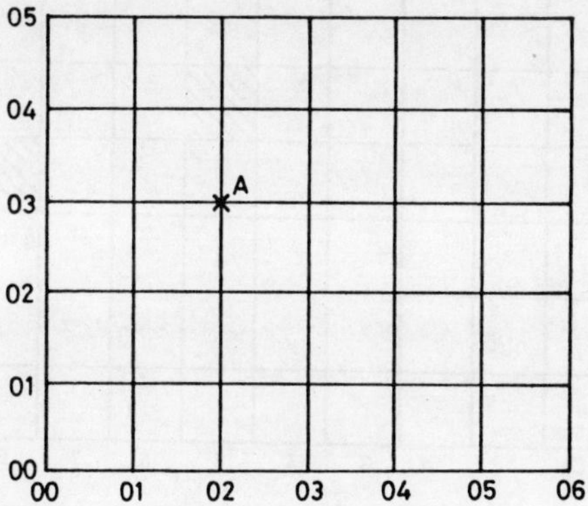

The coordinates of A would be (02, 03). To read from this mesh, it means that a four digit random number table would be required but the two digit table can also be converted into four digit table by combining two columns together.

The coordinates can be found with the help of two digit table if the mesh is drawn thus:

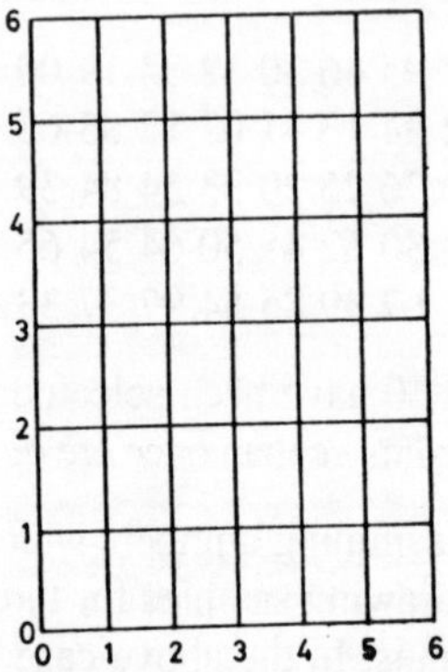

To draw the samples, when the area is to be taken into consideration, a mesh or grid is to be drawn on the map on a proper scale. The boxes may be squares or rectangles. Instead of assigning numbers to the lines as was done in previous case, the columns are numbered thus:

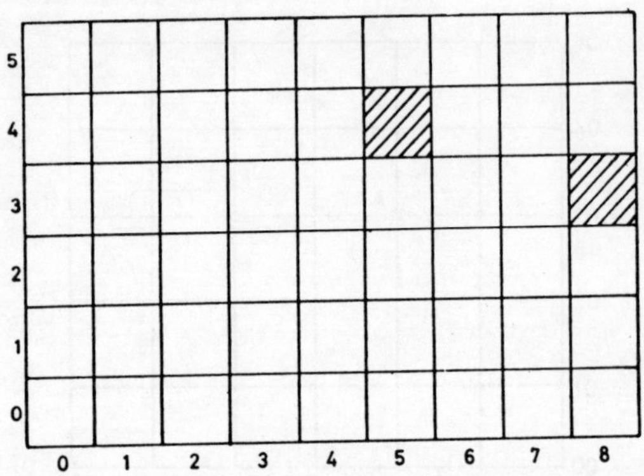

The coordinates of the box A are 5,4 and that of B are 8,3.

Illustration:

For the survey of the fields of a village, 12 per cent samples are to be drawn using mesh with columns numbered.

First of all a mesh has to be prepared on the tracing paper so as to cover all the fields of the village.

The mesh on the tracing paper:

8						85	86		
7									
6			62						
5									
4				43			46		48
3		31						37	
2						25			
1									
0								07	
	0	1	2	3	4	5	6	7	8

A two digit table may be used in this case. The above mesh covers the entire village which has 81 boxes. Twelve per cent sample means 10 boxes. Since there are 9 columns in northings and 9 in eastings, from the samples drawn, 9 has to be ignored.

Starting from anywhere in the random sampling number tables say from 07, (row 11, column 4) the following samples are drawn:

07 90 86 49 85 79 48 25 43 96 46 62 31 37

In the above samples, 90 49 79 and 96 have to be discarded because there are only 9 columns. Mark in the boxes the samples that have been drawn. Place the mesh on the village and survey the fields below these marked boxes.

Illustration:

A village with 300 fields has to be surveyed. The village has four distinct soil zones. The percentages of land under each soil zones are as follows:

Good Quality Soil (A)	40 per cent
Medium Quality Soil (B1)	20 per cent
Medium Quality Soil (B2)	25 per cent
Poor Quality Soil (C)	15 per cent

Using stratified random sampling technique, draw 20 per cent samples. (See Figure 15).

There may be patches of different soil zones spreading here and there in the village. Irrespective of their position, assign numbers to the fields starting from any point.

Soil Zone	Percentage	Size of Samples
A	40	$\frac{40 \times 60}{100} = 24$
Bl	20	$\frac{20 \times 60}{100} = 12$
B2	25	$\frac{25 \times 60}{100} = 15$
C	15	$\frac{15 \times 60}{100} = 9$

It means that 24 samples have to be drawn from the soil zone (A), 12 from the soil zone (B1), 15 from the soil zone (B2) and 9 from the soil zone (C).

Since the total number of fields are 300 which is a 3-digital figure, the table of random sampling numbers would be read in a way such that 3 digits are obtained e.g. we may begin with 201, 774, 947 or we may take 017, 449, 470 or any such combination that gives us a 3-digital figure.

Taking first of all the cases of soil zone (A) from which 24 fields are to be selected. In this group we have fields numbering from 1 to 40 and then from 165 to 175 and then from 251 to 260. The samples drawn beyond these numbers will be discarded.

Let us begin with the first three columns of the table of random sampling numbers. We obtain the following field numbers:

201 744 947 221 932 450 449 162 045 327 036

The number 036 is the only number drawn uptill now that falls within the range of the soil zone (A). In this way 23 other such fields have to be selected which are

017	(017)	172	026	010	(026)	174	009	039
002	(017)	(026)	(009)	260	(009)	(260)	(009)	016
021	259	256	167	170	001	030	(036)	(172)
007	018	040	(259)	(007)	175	(030)	166	033

The figures in brackets have been discarded because they are being repeated.

In a similar manner, the fields under the four other soil zones may be selected and the survey started.

FIG.15. SOIL MAP OF VILLAGE

CHAPTER VII

HYPOTHESIS TESTING

Geography is a science of comparisons between observed and expected data. These comparisons, similarities, associations or relationships between two aspects may be the result of certain group of factors. To identify these factors there may be two such main groups. The first group includes Consistent factors in which there are such factors that predictions on their basis is possible e.g. if the farming systems of arid zones and humid zones are compared, a number of factors are involved. The regime of rainfall is a consistent factor because on its basis predictions can be made which are acceptable.

The other group, on the other hand includes Irregular factor i.e. the factors which operate in such a way that reasonable prediction on their basis cannot be done e.g. in the arid and humid zones farming comparison in which a farmer may deviate from normal factors and may start another type of farming.

In the testing of a hypothesis it has to be seen as to which of these groups operate. When further analysis is made with stronger logical confidence, one must have certain procedure to asses this confidence. It is done by subjective judgement i.e. how far a certain factor can be applicable. It is found by sample drawing in which random sample drawing is the most common and which also has to be tested.

Suppose in a town A, exactly 20 per cent population is of traders. If a sample of 100 is drawn it is necessarily not likely to come out 20 as far as traders population is concerned. It is due to the fact that an irregular factor is operating which is 'chance' and therefore basic prediction cannot be made with great degree of acceptability.

Let there be two towns A and B. The town A has 20 per cent traders and the town B has 10 per cent traders. A sample of 100 is drawn and suppose in town A it comes out to be 20 and in B, 10 which is entirely a matter of chance. It will not be necessarily so every time because of chance factor in drawing sample.

There can be two statements:

1. The 20 per cent occurence of traders population is not a chance occurence but a definite one.

2. The 20 per cent occurence of traders population is a chance occurence and therefore it is not a true representative of the entire population.

Now it has to be inquired that if the percentage of traders population in A and B towns differs or not.? If they differ, it has to be seen that the difference is very great or not ? It has also to be seen that the percentage of population in town A is greater than that in town B or vice versa.

Let the traders population in town A be represented by T_A and that in town B by T_B .

It is to be seen if

$T_A > T_B$ or $T_B > T_A$ or that T_A is not equal to T_B.

To find the answer, a statistical procedure has to be adopted using the method of hypothesis.

Hypothesis 1 $T_A > T_B$ (T_A is greater than T_B)

Hypothesis 2 $T_B > T_A$ (T_B is greater than T_A)

Hypothesis 3 $T_A \neq T_B$ (T_A is not equal to T_B)

There are two types of hypothesis:

1. Null Hypothesis (H_0)
2. Alternative Hypothesis (H_1)

The formulation of H_O and H_1 is very important and it has to be done very carefully. H_O is formulated in negative style and is always initially supposed to be rejected e.g. in Hypothesis 3 above the Null hypothesis that the differences in percentages of traders population in the town A and B are non-existent. They are due to chance occurence therefore the two samples are not acceptable.

The Alternative type of hypothesis may be of two types:

1. Directional and
2. Non- directional

The statement that the percentages of traders population in the town A is greater than that in town B is a directional type of alternative hypothesis whereas the statement that there exists a difference between the percentages of traders population in the town A and B is a non-directional type of alternative hypothesis because in the later statement the direction is not defined. These two types of hypotheses are also very important. A Null hypothesis if rejected as a directional hypothesis may be accepted as a non-directional hypothesis.

Before taking up the problem, first of all the rejection level or the confidence level has to be decided. In some examinations, 33 per cent are the passing marks whereas in some they are 35 per cent i.e. a student securing marks below this level is deemed to have failed or rejected. There is no mathematical proof of these percentages but it has been decided like that.

The rejection level may be taken as 0.05 i.e. 95 per cent acceptability or it may be taken as 0.01 or 99 per cent acceptability.

The distribution of data may be such that 68 per cent variates lie between 1 to –1 standard deviation range or 95 per cent data lies between 2 to –2 standard deviation range or 99 per cent data lies between 3 to –3 standard deviation range. After watching this distribution only, it has to be decided as to what the rejection level should be taken and which test would be most appropriate.

There are the following three main criterias on the basis of which the test is to be selected:

1. Characteristics of the data in the distribution

2. Cost-effectiveness of different tests and

3. Assumptions about the background population from which the sample has to be drawn.

1. Characteristics of the data in the distribution

The characteristics emanate from the type of level in the data. The data may be of nominal, ordinal or interval type i.e. the data with which we are dealing may be expressed in nominal, ordinal or interval scale in accordance with the test that has to be selected. Nominal scale is mainly a count and not a measurement. Ordinal level of measurement is a measure which gives numerical values which may be expressed in absolute or ratio scale. If the data is in the form of percentages, chi-squared test cannot be used whereas if the data is in the nominal, interval or even ordinal form without ratio, chi-squared test can be applied.

2. Cost-effectiveness of different tests

Effectiveness is the ability of the method to give results to a certain degree in which time and energy is required. In statistics, effectiveness has often been referred to as power. If more powerful results are required, the sample size may be smaller whereas relatively less powerful tests require larger samples. Therefore if the investigation that is being carried on has a sample data, more powerful test must be selected. If the sample size is large, the freedom is greater in the selection of test. It may be less powerful.

3. Assumption about the background population from which sample has to be drawn

Before the First World War the practice by the statisticians was that they assumed the background population to be normally distributed. The tests which are suitable for that population which is assumed to be normally distributed are called Parametric Tests.

Sampling is essential because otherwise too much of time and money will be involved. Suppose a survey is to be done of a slope and sampling is not done thinking that the points which make the slope are countable. It is true that the points are countable but the time involved in counting will be very great and so sampling is unavoidable.

Suppose there are 10,000,000 fields which are supposed to be normally distributed. From these fields a sample of 10,000 is taken assuming that it is true representative of the total population. In this case only those tests can be used which are applicable in those cases where the data is normally distributed e.g. if the data has binomial distribution. It leads to a type of test called Parametric Test because in it, it has been assumed that the data is normally distributed.

While testing the hypothesis, the following steps should be followed:

Step 1. Proper formation of Null Hypothesis (H_0) with anticipation of it being rejected.

Step 2. Proper formation of Alternative Hypothesis (H_1).

Step 3. Decision about the rejection level (α)

Step 4. Index of the test i.e. chi-squared test, variance test etc.

Step 5. Conclusion: If the probability is greater than the rejection level (α), the Null Hypothesis (H_O) is rejected and vice versa.

Chi-squared Test

Chi-squared test is a non-parametric test also. It is not applicable if the data is in ratio scale. In this test the distribution is tested against a supposed distribution but generally the average is taken as a supposed distribution. The rejection level of the hypothesis is entirely the choice of the investigator. Chi-squared test has a very wide dimension and it is very similar to the standard deviation. In this test, the Expected Value (E) may be derived from various sources e.g. from the theory of probability of permutation and combination or it may be derived from some well known theoritical distribution e.g. Poisson's distribution or it may also be derived from the hypothesis of inter dependence.

Chi-squared test is tried in various situations to solve varied problems e.g. testing the validity of hypothesis, for determining the randomness of sample, for determining the reliability etc. There are quite a long range of uses depending upon the surveyor. The basic idea is to examine statistics i.e. the data pertaining to certain aspects and in it, observe variations within the same and note some average determinants out of which, by our own precision, select any one as a very significant and predominant factor. After that the frequency of its occurence of significant factors is noted and then variations are related—it may be rejected or accepted.

In chi-squared test it is also possible that the data does not fit into a normal frequency curve and the data possesses marked skewness due to which it is not advisable to transform the data into approximate normal frequency distribution pattern. In parametric tests, first of all mean is to be found out to observe as to how much the values deviate from the mean i.e. it is arranged around the normal curve. In variance test, absolute data is used in the form of percentages or groups but when the data is in the form of percentages, the mean cannot be found.

The relationship between the variates is observed or computed in the form of variance or in the form of standard deviation when the total sum of the standard deviation or variance is taken and the weightage appropriated against the frequency of occurence of the determinant and integrated thus obtaining final weightage index. This index is posted upon the chi-squared graph or read on a chi-squared table to obtain the probability of the correctness of the null hypothesis. The confidence level of 95 per cent and over is sufficient to prove the validity of the null hypothesis. If the acceptence level as decided earlier falls below 95 per cent the null hypothesis is doubtful and if it is below 90 per cent, the null hypothesis is rejected. The null hypothesis should be framed very carefully and well tested. If it is rejected, other determinants are to be found out so that an inverse hypothesis comes out.

Variate	Observed Frequency (O)	Percentage to Total Population	Expected Frequency (E)
x_1	f_1	d_1	f'_1
x_2	f_2	d_2	f'_2
x_3	f_3	d_3	f'_3
x_4	f_4	d_4	f'_4
x_5	f_5	d_5	f'_5
x_6	f_6	d_6	f'_6
	Total = $\sum_{i=1}^{6} f_i$	Total = 100	

If there exists any relationship between the observed frequency and any determinant d_1 d_2 d_3 ...d_6 it must be in the form of percentage and not in terms of actual numbers. If f_1 and d_1 have the same proportion i.e. maximum f_1 has maximum d_1 value and if it is maintained throughout, then d is a significant factor. If the relationship is very intimate, then f and d values will be having the same proportion. The chi–squared value will be

$$\chi^2 = \frac{(f_1 - f_1')^2}{f_1'} + \frac{(f_2 - f_2')^2}{f_2'} + \frac{(f_3 - f_3')^2}{f_3'} + \ldots\ldots + \frac{(f_6 - f_6')^2}{f_6'}$$

$$= \sum_{i=1}^{6} \frac{(f_i - f_i')^2}{f_i'}$$

$$= \sum \frac{(0 - E)^2}{E}$$

where 0 is observed value, E is the expected value and

$$f_i' = \frac{d_i \times T}{100}$$

where f'_i is the expected value; i = 1, 2, 3,......6

d is percentage to the total population and T is time sum of observed frequencies.

Illustration:

In a random sample of 200 villages, the distribution of the villages is observed on the basis of altitude and the following results obtained:

Easily Recognised Zones (Metres)	Observed Frequency Distribution 'O'	Percentage Occupied by Each Zone
2000 – 4000	110	35
4000 – 6000	30	30
6000 – 8000	45	20
8000 – 10000	5	10
10000 – 12000	10	5
Total	200	100

Test whether the distribution of the number of villages are directly related to the altitude zones of the area of investigation.

The percentages occupied by each zone do not correspond with the values in O column. It may be due to the altitude or due to certain other conditions of the village site. Let us therefore first of all find out the expected percentages corresponding with the percentages occupied by each zone. These percentages must be calculated against the total of the observed frequency distribution e.g.

Out of 100, the percentage in 2000 – 4000 metre range is 35

So out of 200, the percentage should be $35 \times 2 = 70$.

Similarly out of 100, the percentage in 4000 – 6000 metre range is 30.

Therefore out of 200, the percentage should be $30 \times 2 = 60$ and so on.

These are the Expected Values (E).

	2000-4000	4000-6000	6000-8000	8000-10000	10000-12000
Observed Value (0)	110	30	45	5	10
Expected Values (E)	70	60	40	20	10
(0 – E)	= +40	–30	+5	–15	+0
$(0-E)^2$	=1600	900	25	225	0
$\frac{(0-E)^2}{E}$	=1600/70	900/60	25/40	225/20	0/10
	= 22.86	15.00	0.63	11.25	0.00

$\therefore \chi^2 = 22.86 + 15.00 + 0.63 + 11.25 + 0.00$

$= 49.74$

Since there are 5 zones, the degree of freedom = 5 – 1 = 4. The chi–squared value of 49.74 when plotted on the chi–squared graph against 4 degree of freedom, lies in the region of less than 0.1 per cent acceptability. It means that the null hypothesis that the distribution of the number of villages is directly related to the altitude zones of the area of investigation is rejected and the reverse is correct upto 99 per cent confidence level. It shows that the altitude is decidedly not a determinant as far as the number of villages are concerned and that some other factors are responsible.

Exercise:

Over an area of different soil types, a sample survey of fields is made and these fields are grouped into the following four categories with the number of fields and the area of the soil zones:

Quality of Soil		Observed Values (0)	Area of Soil Zones (Percentages)
High Fertility Soil	(A_1)	20	10
Medium Fertility Soil	(A_2)	80	45
Medium Fertility Soil	(A_3)	35	20
Low Fertility Soil	(A_4)	15	25
	Total	150	100

Test whether the distribution of the fields is related to the area of various soil zones.

In 100, the area of soil zone A_1 occupies 10 fields so in 150 the expected value for the same soil zone

$$= \frac{10 \times 150}{100} = 15$$

Similarly the expected values for other zones of the quality of soil can be calculated which will be as follows:

For A_2 soil zone the expected value will be 67.5,

For A_3 soil zone it will be 30 and

For A_4 soil zone it will be 37.5.

	A_1	A_2	A_3	A_4
Observed Value 'O'	20	80	35	15
Expected Value 'E'	15	67.5	30	37.5

$(O - E) =$	+5	+12.5	+5	−22.5
$(O - E)^2 =$	25	156.25	25	506.25
$\frac{(O - E)^2}{E} =$	$\frac{25}{15}$	$\frac{156.25}{67.5}$	$\frac{25}{30}$	$\frac{506.25}{37.5}$
	= 1.67	2.31	0.83	13.5

$$\chi^2 = \sum \frac{(O - E)^2}{E} = 1.67 + 2.31 + 0.83 + 13.5$$

$$= 18.31$$

Degree of Freedom $= 4 - 1 = 3$

When the chi–squared value of 18.31 is plotted against the degree of freedom of 3 on the chi–squared graph, it is observed that the confidence level is less than 0.1. per cent. The null hypothesis is therefore rejected i.e. the distribution of the fields is not related to the areas of the various soil zones. Therefore some other factors may be tried.

Variance Analysis

Chi-squared test is an indirect approach of testing any hypothesis because when the test fails, it means that the opposite is correct which is an indirect method. Instead of indirect form, direct assessment can be made which is based on variance analysis. It provides an objective assessment of the validity of the differences between the two mean values.

Suppose the crop yield pattern is to be studied in an area which has three soil zones. The area of the region and the number of farms contained within is so large that the study of individual farm is not possible or practicable. Large number of fields necessitate sampling and therefore a sample of 10 fields from each of the 3 soil zones is drawn and the yield of crops in all the thirty fields which constitute the over all sample population is recorded. The problem is to estimate as to whether the differences in the crop yields noted are significantly related to the soil types or they are a result of the random sample drawn. If the differences between the various sets of sample are not significant, it can be confidently stated that the differences in the yields are not due to the differences in the soil types whereas on the contrary, if the differences of yields between the sets of samples are significant, it strongly indicates

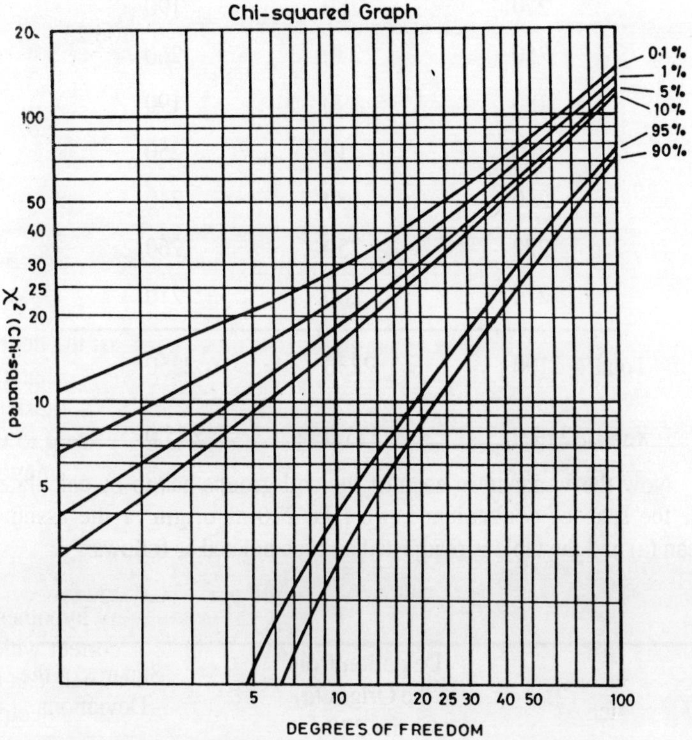

Based on data in D.V. Lindley and J.C.P. Miller, Cambridge Elementary Statistical Tables, Table 3

that the yield differences are directly related to the differences in the soil types.

Soil-wise Yield of Foodgrains
(lbs. per acre)

S_1	S_2	S_3
240	190	210
170	240	290

	190	260	250
	270	280	190
	250	230	260
	180	190	190
	210	210	250
	240	180	240
	220	250	260
	220	200	210
Total	2190	2230	2350
$\bar{x}$ =	219.0	223.0	235.0

Now the variance in each of the soil groups has to be calculated. For the sake of convenience, we take 220 as origin or the assumed mean (a) and the table is readjusted and retabulated as follows:

			Deviation from the Origin (a) i.e.220			Squares of the Deviations		
S_1	S_2	S_3	S_1	S_2	S_3	S_1	S_2	S_3
240	190	210	+20	−30	−10	400	900	100
170	240	290	−50	+20	+70	2500	400	4900
190	260	250	−30	+40	+30	900	1600	900
270	280	190	+50	+60	−30	2500	3600	900
250	230	260	+30	+10	+40	900	100	1600
180	190	190	−40	−30	−30	1600	900	900
210	210	250	−10	−10	+30	100	100	900
240	180	240	+20	−40	+20	400	1600	400

220	250	260	00	+30	+ 40	000	900	1600
220	200	210	00	−20	−10	000	400	100
2190	2230	2350	−10	+30	+150	9300	10500	12300

The sum of the deviations from the origin or assumed mean 'a' (220) = 170.

The sum of the squared deviations = 32100.

Correction Factor = $(170)^2/30 = 28900/30 = 963.33$ which is to be redistributed to calculate the variance. If the mean of S_1, S_2 and S_3 would have been 220, the redistribution was not needed but since it is not so, the correction factor has to be applied to calculate the variance.

Squared deviations for r variates from the assumed mean 'a' is $r(x-a)^2$ where x are individual values of S_i when i = 1, 2, 3 or it can be expressed as

$$= r\,(x-a)^2$$

$$= \frac{r^2\,(x-a)^2}{r}$$

$$= \frac{1}{r}[r\,(x-a)]^2$$

$$= \frac{1}{r}[(x-a)+(x-a)+(x-a)+\ldots\ldots + r \text{ times}]^2$$

$$= \frac{1}{r}\left[\sum (x-a)\right]^2$$

Now substituting the values in $\frac{1}{r}[\Sigma(X-a)]^2$, we get

$$= \frac{(-10)^2}{10} + \frac{(+30)^2}{10} + \frac{(+150)^2}{10}$$

$$= \frac{100}{10} + \frac{900}{10} + \frac{22500}{10}$$

$$= \frac{23500}{10}$$

= 2350 which is the over all deviation including the correction factor.

Variance estimate between the sample

Sum of squares	2350.00
Correction factor	963.33
	1386.67

Degree of freedom = 3 – 1 = 2

Variance estimate between the sample = 1386.67/2

= 693.33

Variance estimate within the sample

Total sum of squares	32100.00
Correction factor	963.33
	31136.67
Sum of squares between the sample without correction factor	1386.67
	29750.00

Degree of freedom = 30–1–2

= 27

Variance estimate within the sample = 29750.00/27

= 1101.85

Variance estimates have two components:

a) Between the sample i.e. S_1, S_2, S_3 (Soil type);

b) Within the sample i.e. 30 field samples (10 in each soil type).

For finding out variance estimate, between the sample figures are to be divided by the degree of freedom of 2 and within the sample figures by 27 i.e. there is no common denominator. Therefore direct comparison is not possible and hence variance estimate test ratio or Snedecor's F distribution has to be found out by dividing the greater variance estimate by lesser variance estimate i.e.

$$\text{Snedecor's F distribution} = \frac{\text{Greater Variance Estimate (G.V.E)}}{\text{Lesser Variance Estimate (L.V.E)}}$$

$$= \frac{1101.85}{693.33}$$

In this case the G.V.E. has a degree of freedom of 27 and the L.V. E. has a degree of freedom of 2.

In Table II of percentage point of the F distribution table showing 1% of variance ratio, if the reading is taken for 27 degree of freedom of greater variance estimate corresponding to 2 degree of freedom of lesser variance estimate, it comes as 99. But our value of F is only 1.59. Had it been 99, then it would be a chance difference of 1 in 100 or a probability of factor under consideration as a significant factor with a confidence level of 99%. But since it is not so, the confidence level is below 99%.

Again in the same way when the readings are taken in Table I of 5% level of variance ratio, the reading comes between 19.4 and 19.5. It means that the confidence level of soil being a significant factor is lower than 95% meaning thereby that the statement that the soil being a significant factor cannot be solidly defended with the help of F distribution table. A number of other factors besides soil e.g. irrigation facilities, fertilizers, seeds etc. may be significant factors in controlling the yield. If there is a significant difference between the two components, then it cannot be simply a matter of chance but it is due to some significant factor. If the deviation due to the difference between the samples is not significant and comparable to deviation, it means that the obtained differences are due to chance selection i.e. these differences are not due to any significant factor.

The two components are:

a) The differences in the yields in the same set and hence the deviation;

b) The differences in the yields in separate sets of S_1 , S_2, and S_3.

If S_1, S_2 and S_3 are compared on the basis of yields, the differences in the 30 fields may be due to soil type because there are three different soil types involved (Component b). In component a, the deviation in each group is not duc to soil typc because the soil is of the same category in each of the three sets meaning thereby that other factors besides soil are involved.

If a comparison between the yields in S_1 and S_2 does not show any significant difference, it means that the soil is not a significant factor and a great variation shows the involvement of some significant factor in which the soil may be one of the most significant factor.

If two factors are involved, the analysis is termed as bi-factoral analysis and if more than two factors are involved, the analysis is multi-factoral analysis. In principle component analysis the aim is the same but the process is through matrices which is a much sophisticated method in which orthognal transformations are taken and mapped. The factoral analysis and the principle component analysis are almost the same. They differ only for the students of advanced mathematics.

Illustration:

An area contains 200 villages of different sizes situated at different heights above sea level. Outwardly it appears that the size of the villages generally decreases with the increase in altitude. An investigation is therefore carried out to establish as to whether or not the altitude is a significant determinant of the size of the villages. For this purpose a stratified random sample of 24 villages, given below, is drawn. Determine the level of significance of altitude as a causative factor and on the basis of your findings enunciate an acceptable and valid hypothesis.

Sample with size of village	Height above sea level			
	8000 – 10000	6000 – 8000	4000 – 6000	2000 – 4000
1	180	400	460	380
2	290	310	500	500
3	320	360	480	540
4	210	450	360	560
5		150	310	400
6			290	480
7				520
8				430
9				390
Total	1000	1670	2400	4200

Origin or the assumed mean taken as 400.

Deviation from the origin as 400

	8000 – 10000	6000 – 8000	4000 – 6000	2000 – 4000
	– 220	0	+ 60	– 20
	– 110	– 90	+ 100	+ 100
	– 80	– 40	+ 80	+ 140
	– 190	+ 50	– 40	+ 160
		– 250	– 90	0
			– 110	+ 80
				+ 120
				+ 30
				– 10
Total	– 600	– 330	0	+ 600

Grand total = – 600 – 300 + 0 + 600

= – 330

$$\text{Correction factor} = \frac{(-330)^2}{24} = \frac{108900}{24}$$

$$= 4537.5$$

Checking the correctness of the total for calculating the correcton factor.

Actual total in each category of altitude	1000	1670	2400	4200
Total with respect to assumed origin as 400	1600	2000	2400	3600
Difference	– 600	– 330	0	+ 600

Total difference = – 330 which is equal to the sum of deviations from the origin as 400 calculated earlier.

Squares of the deviations from the origin as 400.

8000 – 10000	6000 – 8000	4000 – 6000	2000 – 4000
48400	0	3600	400
12100	8100	10000	10000
6400	1600	6400	19600
36100	2500	1600	25600
	62500	8100	0
		12100	6400
			14400
			900
			100
103000	74700	41800	77400

Grand total = 296900

$$\frac{(x-\bar{x})^2}{n} = \frac{(-600)^2}{4} + \frac{(-330)^2}{5} + \frac{(0)^2}{6} + \frac{(+600)^2}{9}$$

$$= \frac{360000}{4} + \frac{108900}{5} + \frac{0}{6} + \frac{360000}{9}$$

$$= 90000 + 21780 + 0 + 40000$$

$= 151780$ is the over all deviation including the correction factor.

Variance estimate between the sample		Variance estimate within the sample	
Sum of squares	151780.0	Total sum of squares	296900.0
Correction factor	4537.5	Correction factor	4537.5
	147242.5		292362.5

Sum of squares between the sample without correction factor	147242.5
	145120.5

Degree of freedom between the sample = 4 – 1 = 3

Variance estimate = 147242.5/ 3
= 49080.8

Degree of freedom within the sample = 24 – 1 – 3 = 20.

Variance estimate = 145120.0/ 20
= 7256.0

$$\text{Snedecor's F} = \frac{\text{Greater Variance Estimate}}{\text{Lesser Variance Estimate}} = \frac{49080.8}{7256.0} = 6.76$$

Answer: On the F-Distribution table with 5 per cent variance estimate for 3 degree of freedom at greater variance estimate and for 20 degree of freedom at lesser variance estimate the value comes as 3.1 whereas for the same degrees of freedom at 1 per cent variance estimate, the value comes as 4.9. It means that our F value of 6.76 has a variance level even lower than 1 per cent which indicates that the presumption that the size of the village decreases with the increase in altitude has a very high probability. It therefore establishes that the significance level of altitude as a causative factor in determining the size of the village is very high hence the hypothesis is very valid.

Table of Percentage Points of the F-Distribution
Table I

5 Per Cent Level of Variance Ratio

Number of Degree of Freedom of Lesser Variance Estimate	Number of Degree of Freedom of Greater Variance Estimate							
	1	2	3	4	5	10	20	Infinity
1	161	200	216	225	230	242	248	254
2	18.5	19.0	19.2	19.2	19.3	19.4	19.4	19.5

3	10.1	9.6	9.3	9.1	9.0	8.8	8.7	8.5
4	7.7	6.9	6.6	6.4	6.3	6.0	5.8	5.6
5	6.6	5.8	5.4	5.2	5.0	4.7	4.6	4.4
10	5.0	4.1	3.7	3.5	3.3	3.0	2.8	2.5
20	4.3	3.5	3.1	2.9	2.7	2.3	2.1	1.8
Infinity	3.8	3.0	2.6	2.4	2.2	1.8	1.6	1.0

Table II

1 Per Cent Level of Variance Ratio

Number of Degree of Freedom of Lesser Variance Estimate	Number of Degree of Freedom of Greater Variance Estimate							
	1	**2**	**3**	**4**	**5**	**10**	**20**	**Infinity**
1	4052	5000	5403	5625	5764	6056	6192	6366
2	98	99	99	99	99	99	99	99
3	34	31	29	29	28	27	27	26
4	21	18	17	16	16	15	14	13
5	16	13	12	11	11	10	9.6	9
10	10	7.6	6.6	6.0	5.6	4.8	4.4	3.9
20	8.1	5.8	4.9	4.4	4.1	3.4	2.9	2.4
Infinity	6.6	4.6	3.8	3.3	3.0	2.3	1.9	1.0

CHAPTER VIII

CORRELATION AND REGRESSION

CORRELATION

Correlations refer to a linear relationship between two variates or between two characteristics of a single variate. If a single variate is to be taken into consideration having two characteristics and dimensions, one characteristic of the variate is considered as small average and the other as high average in value. With respect to other characteristic, similar grades can be distinguished. i.e. low value average and high value average. Now it is to be determined whether the low value average and the high value average with respect to one characteristic are also associated in the same sequence with respect to the other characteristic. In case there is a perfect association in the sequence i.e. the values in both increases simultaneously, it means that positive correlation exists. If the sequence is not the same, correlation is not perfect. If on the other hand, the sequence of association is reversed i.e. the lowest value of one characteristic is associated with the highest value of the other characterisitc, there is inverse correlation or negative correlation. There may be or there may not be any correlation at all.

Seeds (kgs.)	Production (kgs.)
50	1000
60	1500
70	1800
80	2000

The above table shows that a positive correlation exists because when the amount of seeds are increased, the total production also increases which is in 100 per cent agreement.

Altitude (Metres)	Population
400	2000
300	3200
200	4800
100	13600

In the above table it is observed that when the altitude increases, the population decreases. Here is also a correlation but in the reverse direction i.e. a negative correlation.

If we take a square as a single item of a single individual it has two characteristics—Perimeter and Sides.

Perimeter in this case refers to the sum of all sides of the square whereas the other characteristic is the length of the sides

Length of the sides (Cms.)	Perimeter (Cms.)
2.0	8
4.0	16
2.5	10
6.0	24

The above table shows a positive correlation but of a special type in which both the characteristics increase in the same ratio. When such a relationship is maintained between two variates, the correlation is said to be Perfect Positive. If the correlation is in the reverse order but it maintains the same rate, the correlation is Perfect Negative.

Fertilizers (Kgs.)	Yield per Acre (Kgs.)
20	150
30	200

40	250
50	320

The minimum of 20 kgs. of fertilizers is associated with the smallest yield of 150 kgs. The medium of 30 and 40 kgs. of fertilizer is associated with a medium yield of 200 to 250 kgs. and the maximum of 50 kgs. of fertilizer with a maximum yield of 320 kgs. It seems to be a case of perfect positive correlation which is not so. The ratio between the fertilizer and the yield would be 20 to 30 and 150 to 200 respectively. The ratio in the fertilizers increases by 1.5 times but the ratio in the yield increases by 1.3 times. Again the ratio between the fertilizer and the yield in 30 to 40 and 200 to 250 respectively i.e. if fertilizers are increased by 1.3, the yield increases by 1.25. Meaning thereby that the characteristics are not maintaining the same rate of increase. It means that the correlation is positive but not perfect.

Fertilizer (kgs.)	Yield in Field I (lbs.)	Yield in Field II (lbs.)	Yield in Field III (lbs.)
10	100	100	100
15	120	110	105
20	180	120	112
25	210	140	115

In the above table, the highest value of fertilizer is not associated with the highest value of yield in fields I and II but the results are not 100 per cent perfect because gaps are still wider. In case of field III, the case is very far from perfectness.

An attempt has been made to emphasize that if there are two variates V_1 and V_2 such that for any change in V_1, there is found to be a proportionate change in the variate V_2, then V_1 and V_2 are said to be correlated. This type of correlation is of nature of cause and effect. One is the cause while the other is the effect. Hence one must recognize independent and dependent variates. Consequently the variates which appear to be caused are independent variates and those that effect are dependent variates e.g.

If is the amount of fertilizer used and y is the yield of crop.

If f increases, y also increases. f is the cause and y is the effect. The reverse cannot be true because f does not depend upon the yield. It is expressed as

$$f + \delta f \rightarrow y + \delta y$$

There are two possibilities:

If with the increase in the independent variate there is an increase in the dependent variate, the correlation is positive. If there is a decrease, the correlation is negative.

The assessment of the degree of correlation between two variables is also essential. In the yields of fields I and II it can be visualised that the one is near to perfect but oral visualization is not suitable because of the whole magnitude of variation between the two variates. Therefore a cartographic technique called Scatter Diagram has been introduced which gives a visual impression of correlation but it cannot be a substitute of the index. But this scatter diagram is one of the characteristics of the correlation.

Perimeter Scale	Side Scale
4	1.0
10	2.5
12	3.0
16	4.0
20	5.0

In case of perfect correlation the above two variates when plotted and joined will generate a straight line (Fig. 16). In case of perfect positive correlation, the straight line will be higher in the right side and have an upward slope. In perfect negative correlation the slope will be downwards as shown by a broken line (Figure 16).

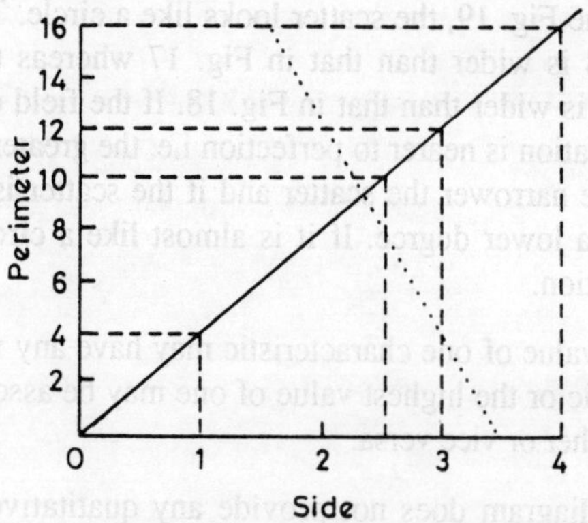

Fig.16. Perfect correlation generating a straight line

If the two items are plotted on a graph in the form of dots in a scatter diagram, and if the points on the outer side are joined by a line, the line will indicate the field of scatter.

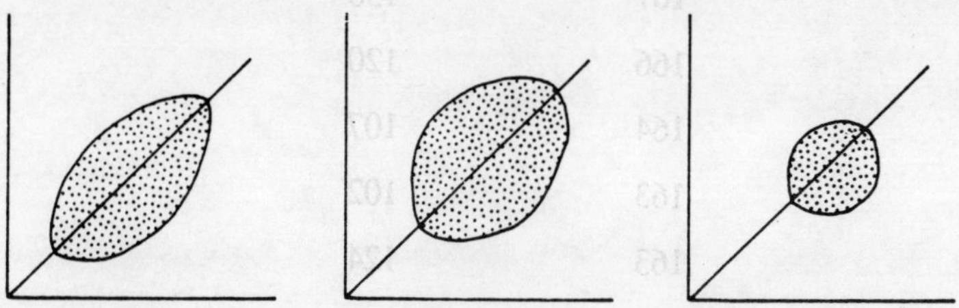
Fig. 17 Fig. 18 Fig. 19

Field of Scatter

In none of the three figures, there is a perfect correlation because perfect correlation generates a straight line i.e. all the points lie on a straight line.

In the three figures the correlation is positive because the upward slope is in the right hand side.

If Fig. 17, the field of scatter is like an ellipse, Fig. 18 is a broader ellipse whereas the Fig. 19, the scatter looks like a circle. The field of scatter in Fig. 18 is wider than that in Fig. 17 whereas the field of scatter in Fig. 19 is wider than that in Fig. 18. If the field of scatter is narrow, the correlation is nearer to perfection i.e. the greater the degree of correlation, the narrower the scatter and if the scatter is wider, the correlation is of a lower degree. If it is almost like a circle, there is almost no correlation.

The highest value of one characteristic may have any value of the other characteristic or the highest value of one may be associated with the lowest of another or vice versa.

The scatter diagram does not provide any quatitative degree of correlationship and unless and until quantitatively assessed, the correlation is not much helpful.

Illustration:

Altitude (metres)	Rainfall (mms.)
167	138
166	120
164	107
163	102
163	124
162	122
161	119
161	123
160	87
160	91
158	99
157	102

156	75
155	69
154	67
153	65
153	64

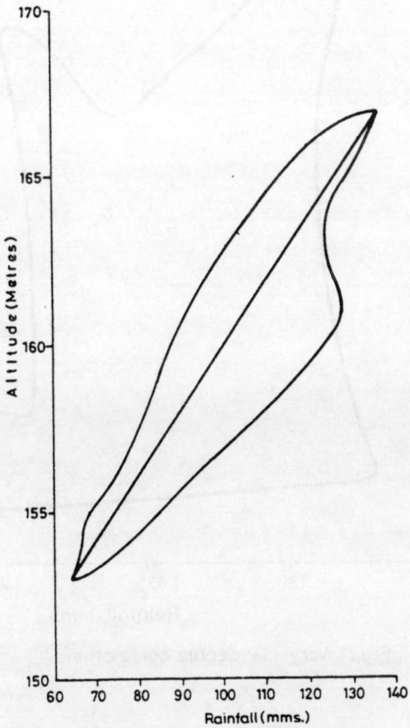

Fig. 20. Partially perfect positive correlation

The scatter diagram Fig 20, prepared with the help of the preceding table shows that the correlation is positive and partially perfect because the line has an upward slope on the right hand side and also the field of scatter is like an elongated ellipse which indicates a higher degree of correlation.

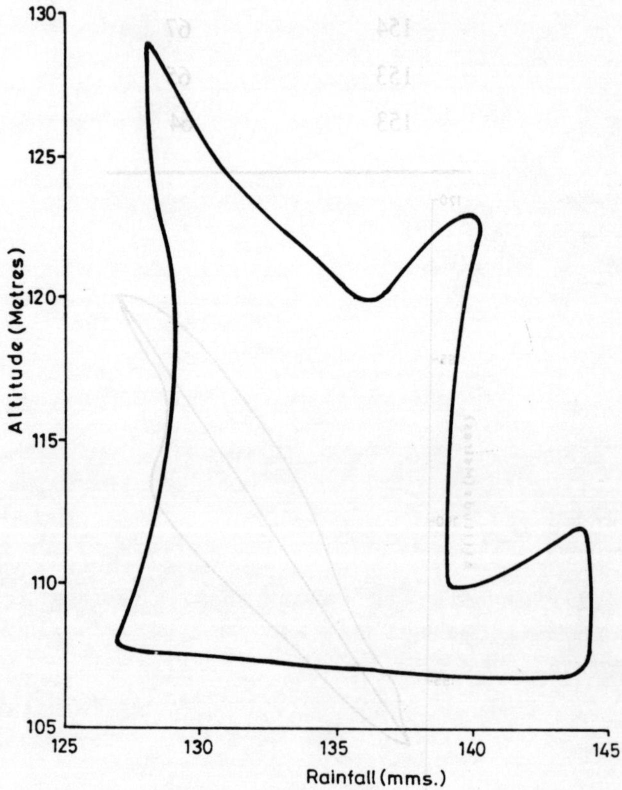

Fig.21. Very low degree correlation

The scatter plotted with the help of the following table in Fig.21 indicates a very low degree of correlation or no correlation at all:

Altitude (Metres)	Rainfall (mms.)
129	128
126	128

123	140
120	136
118	129
115	133
112	144
110	139
108	127
107	144

Before taking up various methods for finding out the Coefficient of Correlation and later on the coefficient of determination one must note the following points very carefully:

1. The Coefficient of Correlation varies between +1 and –1. The correlation is perfect positive when its value comes as +1 and it is perfect negative when it is –1.

2. When the coefficient of correlation has a positive (+) sign there is a positive correlation and when it has a negative (–) sign there is a negative correlation.

3. The degree of correlation decreases as 'r' (Correlation) approaches zero.

There are three main methods for computing the coefficient of correlation:

1. Pearson's Product Moment Correlation Coefficient,

2. Spearman's Rank Correlation Coefficient and

3. Kendall's Rank Correlation Coefficient.

1. Pearson's Product Moment Correlation Coefficient

Pearson gave this method of computing the degree of correlation in terms of numerical index. It is also called Pearson's 'r' or 'r'.

To have a better standing of the technique, let us have an illustration. Let us have 6 sets of a distribution data such that two sets have perfect positive correlation as shown in A and B, and two sets having perfect negative correlation as shown in A and F.

A	B	A×B	C	A×C	D	A×D	E	A×E	F	A×F
18	18	324	12	216	6	108	6	108	0	0
12	12	144	18	216	12	144	0	0	6	72
9	9	81	9	81	18	162	9	81	9	81
6	6	36	0	0	0	0	18	108	12	72
0	0	0	6	0	9	0	12	0	18	0

$\Sigma AB = 585$ $\Sigma AC = 513$ $\Sigma AD = 414$ $\Sigma AE = 297$ $\Sigma AF = 225$

The relationship is to be found with respect to A.

The same values have been tabulated so that the calculation becomes easy and more understandable. The arrangement varies from C. A and B are identical both having higher values in the beginning. In A and F there is perfect negative correlation. C also has a positive correlation but of a smaller degree. D has still smaller degree of correlation.

The Scatter Diagram gives only a pattern whereas 'r' gives a numerical index. ΣAB ΣAC ΣAD ΣAE ΣAF values go on decreasing 585, 513, 414, 297 and 225. It means that at the very outset we can say that since the sum of the products of the individual values go on decreasing, the degree of positive correlation decreases, but the limitation in this system is that it can be used only when the number of items in each pair is the same e.g. 5 in this case. The units of the items also, preferably, be the same. However in case they are not the same, the difficulty could be overcome thus:

	Items	Unit
Suppose A	indicates the marks in a test	Number
B	indicates hours of study	Hours
C	indicates weight of the boy	Kgs.
D	indicates height of the boy	Metres
E	indicates the pocket money	Rupees and so on.

The problem here is that the units are different. When such a situation arises, all the units can be standardised to a common unit with the help of z-score. In this system the entire data is transformed into a new data irrespective of any unit. The formula for this transformation is as follows:

$$\text{z-score} = \frac{x_i - \bar{x}}{\sigma}$$

where x_i are the individual values,

$\bar{x}$ is the arithmetic mean and

σ is the standard deviation.

Taking the case of A we have

Variates (x_i)	Deviation from the mean (d)	d^2
18	18 – 9 = +9	81
12	12 – 9 = +3	9
9	9 – 9 = 0	0
6	6 – 9 = –3	9
0	0 – 9 = –9	81
$\Sigma x_i = 45$		$\Sigma d^2 = 180$

$\bar{x} = 45/5 = 9$ $\qquad \sigma = \sqrt{180/5} = \sqrt{36} = 6$

Substituting the value of x in z for A we get

$$z_{A_{18}} = \frac{18-9}{6} = +9/6 = 1.5$$

$$z_{A_{12}} = \frac{12-9}{6} = +3/6 = 0.5$$

$$z_{A_9} = \frac{9-9}{6} = 0/6 = 0.0$$

$$z_{A6} = \frac{6-9}{6} = -3/6 = -0.5$$

$$z_{A0} = \frac{0-9}{6} = -9/6 = -1.5$$

In this case the arithmetic mean is zero and the standard deviation is 1.

Since the variates in A, B, C, D, E and F are the same the z–score will also be the same but in different order and independent of any unit and therefore A can now be multiplied with B, C, D, E and F

z_A	z_B	z_C	z_D	z_E	z_F
1.5	1.5	0.5	– 0.5	– 0.5	– 1.5
0.5	0.5	1.5	0.5	– 1.5	– 0.5
0.0	0.0	0.0	1.5	0.0	0.0
– 0.5	– 0.5	–1.5	– 1.5	1.5	0.5
– 1.5	– 1.5	–0.5	0.0	0.5	1.5

$z_A \times z_B$	$z_A \times z_C$	$z_A \times z_D$	$z_A \times z_E$	$z_A \times z_F$
2.25	0.75	– 0.75	– 0.75	– 2.25
0.25	0.75	0.25	– 0.75	– 0.25
0.00	0.00	0.00	0.00	0.00
0.25	0.75	0.75	– 0.75	– 0.25
2.25	0.75	0.00	– 0.75	– 2.25
$\Sigma z_A z_B = 5.00$	$\Sigma z_A z_C = 3.00$	$\Sigma z_A z_D = 0.25$	$\Sigma z_A z_E = -3.00$	$\Sigma z_A z_F = -5.00$

$$\frac{\Sigma z_A z_B}{n} = 5/5 = +1.0 \;;\quad \frac{\Sigma z_A z_C}{n} = 3/5 = +0.6. \;;\quad \frac{\Sigma z_A z_D}{n} = .25/5 = +0.05$$

$$\frac{\Sigma z_A z_E}{n} = -3/5 = -0.6; \quad \frac{\Sigma z_A z_F}{n} = -5/5 = -1.0$$

It shows that the coefficient of 'r' varies from +1 to –1.

If r is +1, the correlation is perfect positive.

If r is –1, the correlation is perfect negative.

If it is in between them i.e. zero, there is no correlation at all. From 0 to +1, the values indicate positive correlation and from 0 to –1, the values indicate negative correlation. As the values of positive or negative correlation approach zero, the degree of correlation decreases.

Exercise: With the help of the table given below, calculate the Pearson's Product Moment Correlation Coeffecient.

District	Total Population in Lakhs (x)	Population of Backward Class in Lakhs (y)	$(x-\bar{x})$	$(y-\bar{y})$	$(x-\bar{x}) \times (y-\bar{y})$ +	$(x-\bar{x}) \times (y-\bar{y})$ –
Goalpara	22.25	4.20	7.294	1.381	10.073	
Kamrup	28.54	4.62	13.584	1.801	24.465	
Darrang	17.36	2.62	2.404	– 0.199		0.478
Nowgong	16.81	2.92	1.854	0.101	0.187	
Sibsagar	18.37	2.11	3.414	– 0.709		2.420
Lakhimpur	21.23	3.63	6.274	0.811	5.088	
Mikir Hills	3.79	2.19	– 11.166	– 0.629	7.023	
North Cachar Hills	0.76	0.53	– 14.196	– 2.289	32.495	

Cachar	17.13	2.24	2.174	– 0.579	1.259
Mizo Hills	3.32	3.13	– 11.636	0.311	3.619

$\Sigma x = 149.56$ $\Sigma y = 28.19$ $\Sigma(x-\bar{x})(y-\bar{y}) = 79.331 - 7.776$

$= 71.555$

$\bar{x} = 149.56/10 = 14.956;\quad \bar{y} = 28.19/10 = 2.819$

$$\text{Covariance}\,(x\,y) = \frac{1}{n}\sum(x-\bar{x})(y-\bar{y})$$

$$= 71.555/10 = 7.1555$$

Now the covariance when divided by the product of standard deviations of x and y will give the product moment correlation coefficient 'r'.

$$r = \frac{\frac{1}{n}\sum(x-\bar{x})(y-\bar{y})}{\sigma x\,\sigma y}$$

Now substituting the values we get

$$r = \frac{7.155}{8.742 \times 1.115}$$

$$= 7.155/9.747 = 0.734$$

To avoid too much of calculations both of these formulas for calculating covariance and the standard deviation, can be derived in a much simpler form:

$\frac{\Sigma x}{n} = \bar{x}$ and $\frac{\Sigma \bar{x}}{n}$ is also equal to $\bar{x}$

because $\bar{x} + \bar{x} + \bar{x} + \bar{x} \ldots\ldots$ added n times and divided by $n = \bar{x}$

$$S_{xy}\ \text{(Covariance)} = \frac{1}{n}\sum(x-\bar{x})(y-\bar{y})$$

$$= \frac{1}{n}\left(\sum xy - \sum x\bar{y} - \sum \bar{x}y + \sum \bar{x}\,\bar{y}\right)$$

$$= \frac{\sum xy}{n} - \frac{\sum x\bar{y}}{n} - \frac{\sum \bar{x}y}{n} + \frac{\sum \bar{x}\,\bar{y}}{n}$$

$$= \frac{\sum xy}{n} - \bar{y}\left(\frac{\sum x}{n}\right) - \bar{x}\left(\frac{\sum y}{n}\right) + \frac{\sum \bar{x}\,\bar{y}}{n}$$

Since $\frac{\sum x}{n} = \bar{x}$ and $\frac{\sum y}{n} = \bar{y}$

$$= \frac{\sum xy}{n} - \bar{x}\,\bar{y} - \bar{x}\,\bar{y} + \bar{x}\left(\frac{\sum \bar{y}}{n}\right)$$

$$= \frac{\sum xy}{n} - 2\bar{x}\,\bar{y} + \overline{xy}$$

$$= \frac{\sum xy}{n} - \bar{x}\,\bar{y}$$

$$= \frac{1}{n}\sum xy - \bar{x} \times \bar{y}$$

In the same way the formula for the standard deviation can also be simplified thus:

$$\sigma x = \sqrt{\frac{\sum (x - \bar{x})^2}{n}}$$

$$= \sqrt{\frac{\sum (x - \bar{x})(x - \bar{x})}{n}}$$

$$= \sqrt{\frac{\sum (x^2 - 2x\bar{x} + \bar{x}^2)}{n}}$$

$$= \sqrt{\frac{\sum x^2}{n} - \frac{\sum 2\bar{x}}{n} \times \frac{\sum x}{n} + \frac{\sum \bar{x}^2}{n}}$$

since $\frac{\sum x}{n} = \bar{x}$ and $\frac{\sum \bar{x}}{n}$ is also $= \bar{x}$, we have

$$\sigma x = \sqrt{\frac{\sum x^2}{n} - 2\bar{x} \times \bar{x} + \bar{x}^2}$$

$$= \sqrt{\frac{\sum x^2}{n} - 2\bar{x}^2 + \bar{x}^2}$$

$$= \sqrt{\frac{\sum x^2}{n} - \bar{x}^2}$$

Similarly $\sigma y = \sqrt{\frac{\sum y^2}{n} - \bar{y}^2}$

The formula for $r = \frac{S_{xy}}{\sigma x \sigma y}$ becomes

$$r = \frac{\frac{1}{n}\sum xy - \bar{x} \times \bar{y}}{\sqrt{\left(\frac{\sum x^2}{n} - \bar{x}^2\right)\left(\frac{\sum y^2}{n} - \bar{y}^2\right)}}$$

The above formula can further be transformed in the following form:

$$r = \frac{\sum xy - \frac{\sum x \sum y}{n}}{\sqrt{\left[\sum x^2 - \left(\frac{\sum x}{n}\right)^2\right]\left[\sum y^2 - \left(\frac{\sum y}{n}\right)^2\right]}} \text{ as shown below}$$

$$r = \frac{\frac{1}{n}\sum xy - \bar{x} \times \bar{y}}{\left(\frac{\sum x^2}{n} - \bar{x}^2\right)\left(\frac{\sum y^2}{n} - \bar{y}^2\right)}$$

we have $\bar{x}^2 = \left(\frac{\sum x}{n}\right)^2 = \frac{(\sum x)^2}{n^2}$

$$\therefore r = \frac{\frac{\sum xy}{n} - \bar{x}\,\bar{y}}{\sqrt{\left(\frac{\sum x^2}{n} - \frac{(\sum x)^2}{n^2}\right)\left(\frac{\sum y^2}{n} - \frac{(\sum y)^2}{n^2}\right)}}$$

$$= \frac{\frac{\Sigma xy}{n} - \bar{x}\,\bar{y}}{\sqrt{\left[\frac{1}{n}\left(\Sigma x^2 - \frac{(\Sigma x)^2}{n}\right)\right] \times \left[\frac{1}{n}\left(\Sigma y^2 - \frac{(\Sigma y)^2}{n}\right)\right]}}$$

$$= \frac{\frac{\sum xy}{n} - \bar{x}\,\bar{y}}{\sqrt{\frac{1}{n^2}\left(\sum x^2 - \frac{(\sum x)^2}{n}\right)\left(\sum y^2 - \frac{(\sum y)^2}{n}\right)}}$$

$$= \frac{\frac{\sum xy}{n} - \bar{x}\,\bar{y}}{\frac{1}{n}\sqrt{\left(\sum x^2 - \frac{\left(\sum x\right)^2}{n}\right)\left(\sum y^2 - \frac{\left(\sum y\right)^2}{n}\right)}}$$

$$= \frac{n\left(\frac{\sum xy}{n} - \bar{x}\,\bar{y}\right)}{\sqrt{\left(\sum x^2 - \frac{\left(\sum x\right)^2}{n}\right)\left(\sum y^2 - \frac{\left(\sum y\right)^2}{n}\right)}}$$

since $\bar{x} = \sum x/_n$ and $\bar{y} = \sum y/_n$ we have

$$= \frac{n \times \frac{\sum xy}{n} - n \times \frac{\sum x}{n} \times \frac{\sum y}{n}}{\sqrt{\left(\sum x^2 - \frac{\left(\sum x\right)^2}{n}\right)\left(\sum y^2 - \frac{\left(\sum y\right)^2}{n}\right)}}$$

$$= \frac{\sum xy - \frac{\sum x \times \sum y}{n}}{\sqrt{\left(\sum x^2 - \frac{\left(\sum x\right)^2}{n}\right)\left(\sum y^2 - \frac{\left(\sum y\right)^2}{n}\right)}} \qquad (1)$$

or multiplying the numerator and the denominator in (1) by n we get another form

$$r = \frac{n\sum xy - \sum x \sum y}{\sqrt{\left(n\sum x^2 - \left(\sum x\right)^2\right)\left(n\sum y^2 - \left(\sum y\right)^2\right)}}$$

substituting the values in the preceding table we get

$$r = \frac{493.16 - (149.56)(28.19)/10}{\sqrt{(3001.11 - 22368.19/_{10})(91.90 - 794.68/_{10})}}$$

$$= \frac{493.16 - 421.61}{\sqrt{(3001.11 - 2236.819)(91.90 - 79.468)}}$$

$$= \frac{71.55}{\sqrt{(764.29)(12.43)}}$$

$$= \frac{71.55}{\sqrt{9500.12}}$$

$$= 71.55/97.46$$

$$= 0.734$$

2. Spearman's Rank Correlation Coefficient

Spearman's Rank Correlation Coefficient is expressed as 'r_s'. This method is of a special significance when the data is given only in the form of ranks. If the data is in the form of actual values, then too it can be converted in the form of ranks.

Illustration:

District	Total Population in Lakhs (A)	Population of Backward Class (B)
Goalpara	22.25	4.20
Kamrup	28.54	4.62
Darrang	17.36	2.62
Nowgong	16.81	2.92
Sibsagar	18.37	2.11
Lakhimpur	21.23	3.63
Mikir Hills	3.79	2.19
North Cachar Hills	0.76	0.53
Cachar	17.13	2.24
Mizo Hills	3.32	3.13

First of all the data has to be transformed in ordinal scale as shown below:

Districts	Ranks in (A)	Ranks in (B)	Rank Difference (A) – (B) = d	d^2
Goalpara	2	2	2 – 2 = 0	0
Kamrup	1	1	1 – 1 = 0	0
Darrang	5	6	5 – 6 = –1	1
Nowgong	7	5	7 – 5 = +2	4
Sibsagar	4	9	4 – 9 = –5	25
Lakhimpur	3	3	3 – 3 = 0	0
Mikir Hills	8	8	8 – 8 = 0	0
North Cachar Hills	10	10	10 – 10 = 0	0
Cachar	6	7	6 – 7 = –1	1
Mizo Hills	9	4	9 – 4 = +5	25
				$\Sigma d^2 = 56$

According to Spearman's Rank Correlation Coefficient we have

$$r_s = 1 - \frac{6\Sigma d^2}{n\left(n^2 - 1\right)}$$

where d is the rank different and n is the number of the items in the data.

Substituting the values of d and n we get

$$r_s = 1 - \frac{6 \times 56}{10\left(10^2 - 1\right)}$$

$$= 1 - \frac{336}{10\left(100 - 1\right)}$$

$$= 1 - \frac{336}{10 \times 99}$$

$$= 1 - \frac{336}{990}$$

$$= 1 - 0.3394$$

$$= +0.6606 \quad \text{i.e. positive correlation.}$$

Since 0.6606 is nearer to 0.5, it can be said that there is not much correlation between the total population and the population of the backward class.

Another illustration:

Let there be same ranks in both the items.

Region	Fertilizers (kg./hectare) (A)	Yield (B)	Ranks in (A)	Ranks in (B)	(A) – (B) = d	d^2
I	300	700	1	1	0	0
II	175	600	4	4	0	0
III	250	675	2	2	0	0
IV	200	650	3	3	0	0
V	150	570	5	5	0	0
						$\Sigma d^2 = 0$

$$r_s = 1 - \frac{6 \times 0}{5\left(5^2 - 1\right)}$$

$$= 1 - \frac{0}{5 \times 24}$$

$$= 1 - 0$$

$$= +1 \quad \text{i.e. perfect positive correlation}$$

Let us see the result when there is a small variation.

Region	Fertilizers	Yield (B)	Ranks in (A)	Ranks in (B)	Rank Difference	d^2
I	300	570	1	5	– 4	16
II	175	650	4	3	+ 1	1
III	250	675	2	2	0	0
IV	200	600	3	4	– 1	1
V	150	700	5	1	+ 4	16
						$\Sigma d^2 = 34$

$$r_s = 1 - \frac{6 \times 34}{5(25-1)}$$

$$= 1 - \frac{204}{120}$$

$$= 1 - 1.7$$

$= -0.7$ i.e. somewhat near perfect negative correlation.

However, if the sequence in the ranks in the two variates is in the reverse direction i.e. highest rank of A is opposite to lowest rank in B there is perfect negative correclation.

Illustration:

Region	Fertilizers (A)	Yield (B)	Ranks in (A)	Ranks in (B)	Ranks Difference	d^2
I	150	570	5	1	+ 4	16
II	175	600	4	2	+ 2	4
III	200	650	3	3	0	0
IV	250	675	2	4	– 2	4
V	300	700	1	5	– 4	16
						$\Sigma d^2 = 40$

$$r_s = 1 - \frac{6 \times 40}{5(25-1)}$$

$$= 1 - \frac{240}{120}$$

$$= 1 - 2$$

$$= -1$$ i.e. a perfect negative correlation.

Sometimes it so happens that an item occurs more than once in a data. In this case the middle rank is awarded to both the values i.e. the sum of the ranks preceding and following that value which is being repeated and which is then divided by two.

Serial Number	Variates	Ranks
1	2.33	1
2	2.18	2.5
3	2.18	2.5
4	1.86	4
5	1.79	6
6	1.79	6
7	1.79	6

The second and third values are being repeated. The rank preceding the second value is 1 and the one following the third value is 4.

Therefore 1 + 4 = 5/ 2 = 2.5.

So the second and the third value both will be given 2.5th rank. In the same way the last three values are being repeated. The rank of the one preceding the first repeated value is 4 and the one following the third value is 8. Therefore 4 + 8 = 12/2 = 6. The last three variates therefore will be ranked as 6, 6, 6.

The values of the coefficient of correlation deviate between +1 and –1. In case of the first exercise done earlier the value of r was obtained as + 0.6606. But this value of r gives a very vague idea about the actual position of correlation. To make it more purposeful, it is represented in the form of percentage by obtaining the Coefficient of Determination which is equal to r^2. Therefore the exercise in which r was calculated as +0.6606, the Coefficient of Determination = $(0.6606)^2$

Coefficient of determination $= (0.6606)^2$

$= 0.4364$

$= 0.4364 \times 100$

or 43.64 per cent

i.e there is a positive correlation to the extent of 43.64 per cent.

3. Kendall's Rank Correlation Coefficient

In Kendall's method for calculating the coefficient of correlation, out of the two aspects between which the correlation is to be found, the variates in one aspect are arranged in the order of ranking and named Set A. The rankings corresponding to the variates in the Set A are put against them in another column called Set B. Given below is an illustration from the population figures of the district of Assam.

Districts	Total Population (Lakhs) Set A	Population of Backward Class (Lakhs) Set B	Data Set A Rank	Data Set B Rank
Kamrup	28.54	4.62	1	1
Goalpara	22.25	4.28	2	2
Lakhimpur	21.23	3.63	3	3
Sibsagar	18.37	2.11	4	9
Darrang	17.36	2.62	5	6
Cachar	17.13	2.24	6	7
Nowgong	16.81	2.92	7	5
Mikir Hills	3.79	2.19	8	8
Mizo Hills	3.32	3.13	9	4
North Cachar Hills	0.76	0.53	10	10

Now from the Set B, starting from the top, we have to see as to how many ranks are higher downwards from the first, second, third and so on upto 10 figure. In the case of Kamrup e.g. it is +9; in Goalpara +8, in Lakhimpur +7 and so on. These values are put in one column showing the number of higher ranks downwards.

In the same way again starting from the top, we have to see as to how many ranks are lower downwards from the first, second, third and so on upto the 10 figures. In the above table in case of Kamrup, it is 0, in Goalpara 0, in Lakhimpur 0 and so on. These figures are put in another column showing the number of lower ranks downwards. The difference of each row in these two columns is taken and added to give an index S as shown in the following table:

Districts	Data Set A Rank	Data Set B Rank Downwards	Number of Higher Rank Downwards C	Number of Lower Rank D	Total C + D
Kamrup	1	1	+ 9	0	+ 9
Goalpara	2	2	+ 8	0	+ 8
Lakhimpur	3	3	+ 7	0	+ 7
Sibsagar	4	9	+ 1	– 5	– 4
Darrang	5	6	+ 3	– 2	+ 1
Cachar	6	7	+ 2	– 2	0
Nowgong	7	5	+ 2	– 1	+ 1
Mikir Hills	8	8	+ 1	– 1	0
Mizo Hills	9	4	+ 1	0	+ 1
North Cachar Hills	10	10	0	0	0
			$S^+ = 34$	$S^- = 11$	$S = 23$

Now according to Kendall's formula

$$\tau\,(\text{Tau}) = \frac{S}{\frac{1}{2} \times n\,(n-1)}$$

$$= \frac{23}{\frac{1}{2} \times 10\,(10-1)}$$

$$= \frac{23}{\frac{1}{2} \times 10 \times 9}$$

$$= + \frac{23}{45} = + 0.5111$$

This can be calculated the with help of another formula

$$\tau\,(\text{Tau}) = \frac{S^+ - S^-}{S^+ + S^-}$$

$$= \frac{34 - 11}{34 + 11}$$

$$= +23/45 = +0.5111$$

i.e. using Kendall's technique, the coefficient of correlation is +ve the value of which is 0.5111.

Making use of the same population table of the Districts of Assam, when the coefficient of correlation was calculated with the help of Spearman's formula, r came out to be +0.6606 which is different from that calculated using Kendall's technique. No doubt the coefficient of correlation varies from +1 to –1, but since the scaling within these limits is not the same, the values of r cannot be compared.

On the basis of the three methods discussed earlier, using a common data, it is found that the value of r is different.

In Spearman's method r comes	= + 0.6606
In Kendall's method it is	= + 0.5111
In Pearson's method it is	= + 0.7340

It has already been said that since the scaling within the limits of +1 and –1 are different, the different values of r cannot be compared. Its further explanation is beyond the scope of this book.

Having gone through the three techniques of the coefficient of correlation, we come to the following conclusions:

r always varies between +1 and –1.

If r is +1, the correlation is perfect +ve.

If r is –1, the correlation is perfect –ve.

If r is zero, it means that there is no correlation at all.

From 0 to +1 are the values which give positive r.

From 0 to –1, the values give negative r.

As r approaches zero, the degree of correlation decreases.

Suppose the value of r comes as 1/ 3 or 0.33 in case A and 1/6 or 0.166 in case B. It means that in case A there is greater correlation as compared to B because 0.166 is nearer to zero. It appears that A has two times higher correlation but actually it is not so. It is not indicated like that. The only thing that can be mentioned is that A has a higher degree of correlation.

The actual degree of correlation is found out with the help of a test called Significance Test.

Coefficient of correlation $= \sqrt{\text{Coefficient of Determination}}$

or Coefficient of Determination $= (r)^2$

In case where the value of r is 1/ 3

Coefficient of Determination $= (1/3)^2$

$= 1/9$

$= .11$ i.e. 11%

When r = 1/6, coefficient of determination $= (1/6)^2$

$= 1/36$

$= .0278$ or 2.78%

It means that where r is 1/ 3, it is about four times more significant than that value when r is 1/6.

The same test can be applied in determining the significance of samples also. The samples cannot be 100 per cent representative of total population. Hence it is never ideal. A sample which is nearer to the total population is always supposed to be better. Samples also have certain errors. Therefore there must be a significance level. If the sample drawn is 1/ 3 of the total population, one must know its level of significance with respect to the entire population.

Let there be 1000 items and a sample of 100 has been drawn. How much is the significance of the sample drawn can be obtained by the coefficient of correlation which is perfectly usable?. Significance level actually indicates the level of its acceptability.

If the chance probability is 1 per cent, it is 99 per cent significant. If the chance probability is 5 per cent, it is 95 per cent significant. If the chånce level is less than 95 per cent, it is generally not acceptable. The acceptability level is tested with the help of a test called Student's Test or 't' test in which

$|t| = \frac{r\sqrt{n-2}}{\sqrt{1-r^2}}$ where r is the coefficient of correlation, n are the number of pairs and n – 2 is the degree of freedom for the coefficient of correlation.

If the calculated value of $|t|$ comes out to be greater than the table at the chance probability (significance level), the value of r may be treated as reliable otherwise not.

Illustration:

Using the value of r in the Product Moment Correlation Coefficient exercise which comes as .734 with a degree of freedom (n–2) i.e.8, we get the value of $|t|$ equal to

$$|t| = \frac{.734\sqrt{10-2}}{\sqrt{1-(.734)^2}}$$

$$= \frac{.734\sqrt{8}}{\sqrt{1-.538756}}$$

$$= \frac{.734 \times 2.828}{\sqrt{.461244}}$$

$$= \frac{2.0757}{.6791}$$

$$= 3.056$$

Since the calculated value of | t | i.e. 3.056 when plotted on the Student's t graph against the degree of freedom equal to 8, lies between the |t| curve of 1 per cent and 5 per cent. Thus the reliability of r is more than 95 per cent but less than 99 per cent.

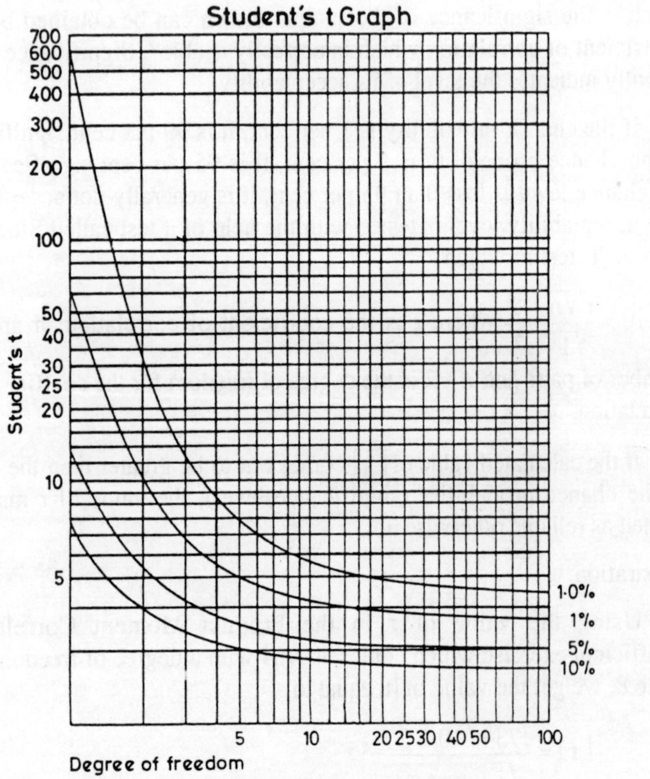

Based on data in D.V. Lindley and J.C.P. Miller, Cambridge Elementary Statistical Tables, Table 3

REGRESSION

Linear Regression for Two Variables

The idea of regression was first of all initiated by Francis. It is generally expected that the children of parents appear to be gifted equally or not equally. These gifts tend to regress both ways—towards the higher side or towards the lower side hence the word regress or regression has been used. The basic theme is that 100 per cent correct prediction about the gifts of children on the basis of the talents of their parents is not possible. However it does not mean to negate the position that it was fairly acceptable and it is acceptable to predict that the children of gifted parents are relatively more gifted than those of ungifted parents. They do not have the same level of gift.

Person	Monthly Salary	Annual Salary
A	Rs 200/-	Rs 2400/-
B	Rs 300/-	Rs 3600/-
C	Rs 400/-	Rs 4800/-
D	Rs 500/-	Rs 6000/-

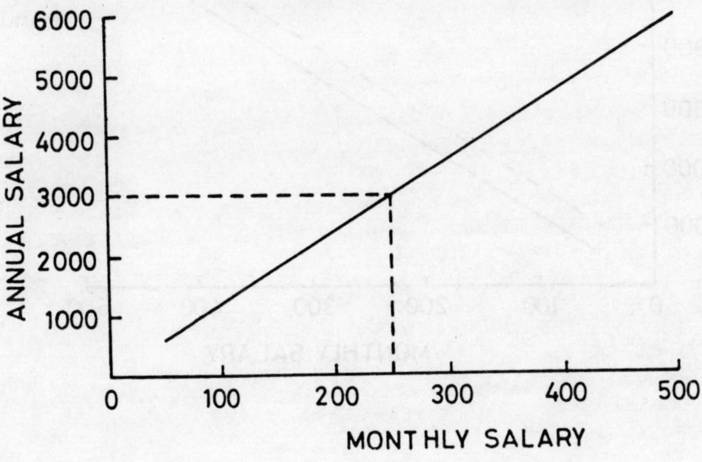

With the help of the graph on page 161, the annual salary of a person drawing Rs 250 per month can be predicted as Rs 3000 and so on, for other monthly salaries also. The annual income increases with the monthly income.

If y is the total salary in x months

a is the salary per month

then y = ax in which y and x are variables and a is constant.

As x changes y also changes.

If a bonus of Rs 500/- is awarded at the end of a year,

A gets Rs 2400/- + Rs 500/- = Rs 2900/-

B gets Rs 3600/- + Rs 500/- = Rs 4100/-

C gets Rs 4800/- + Rs 500/- = Rs 5300/-

D gets Rs 6000/- + Rs 500/- = Rs 6500/-

In A, y = 12(x) + 500 or y = ax +b where b is constant.

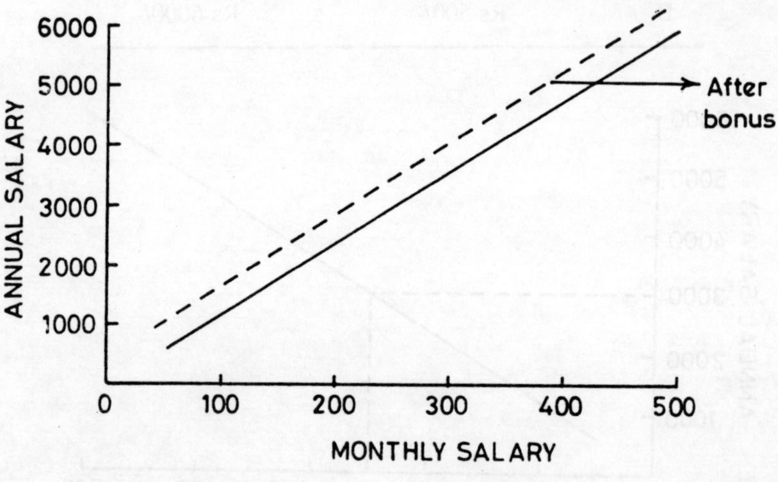

In the graph, on page 162 prediction is possible.

Years	Population
1951	10,000
1961	15,000
1971	25,000

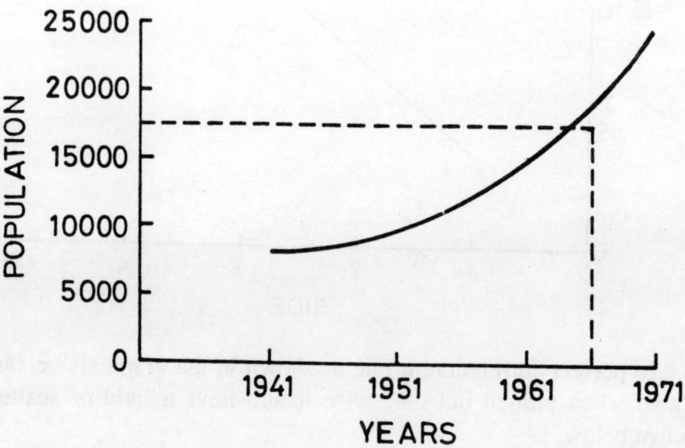

In the above graph, the population of 1965 cannot be predicted because the rate of increase is different and it gradually decreases with the march of time. In the case of salaries, prediction is possible because the rate of increase is constant.

In perfect positive correlation

if the side is 1, the perimeter is 4

if the side is 4, the perimeter is 16

if the side is 2, the perimeter is 8

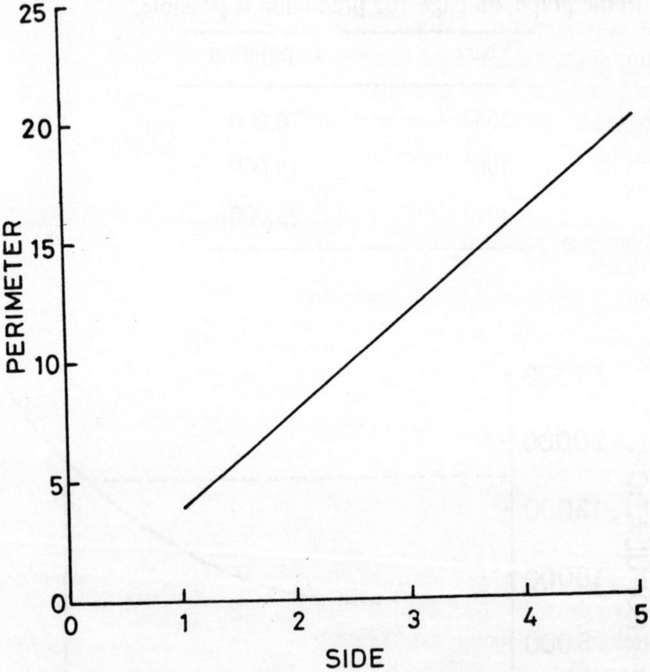

In perfect correlation, a line as shown in the graph above can be drawn when plotted but otherwise it will have a field of scatter as shown below.

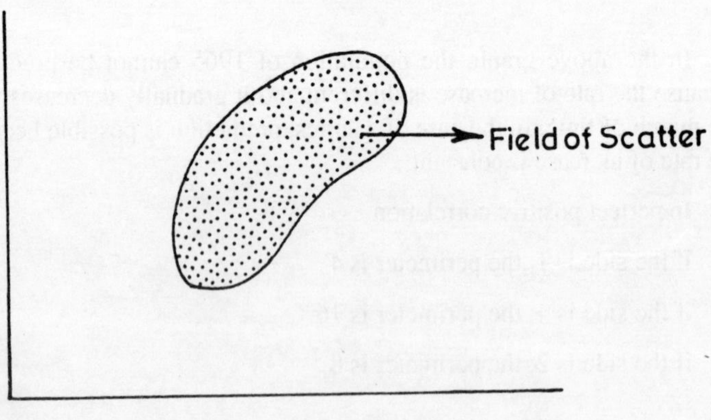

The value of a perfect unknown variate can be found out by graph if its corresponding variate is known. But how to find it in case of a scatter diagram? Take the horizontal and vertical distances of each line from x and y axes and then interchange x axis with y axis and the y axis with x axis. A line has to be found out such that the squares of the distances have minimum deviation i.e. the squares of the vertical distances from the scatter point is the least.

$$y = a + bx, \text{ where b is the base.}$$

To insert the line to be drawn, the angle of inclination must be known. The line may be as shown below

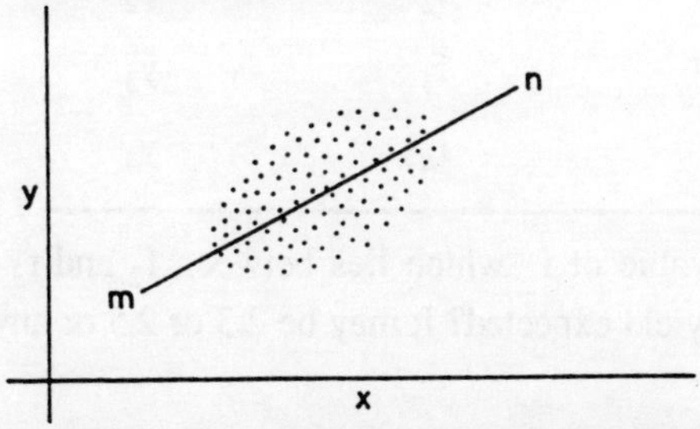

The line should be such that the squares from these points of scatter should be smallest. To determine its location, its slope and the starting point should be known i.e. 'a' and 'b' must be known.

$$b = \frac{n \sum x\, y - \sum x \sum y}{n \sum (x)^2 - (\sum x)^2}$$

Where n are the number of items and x and y are the variates.

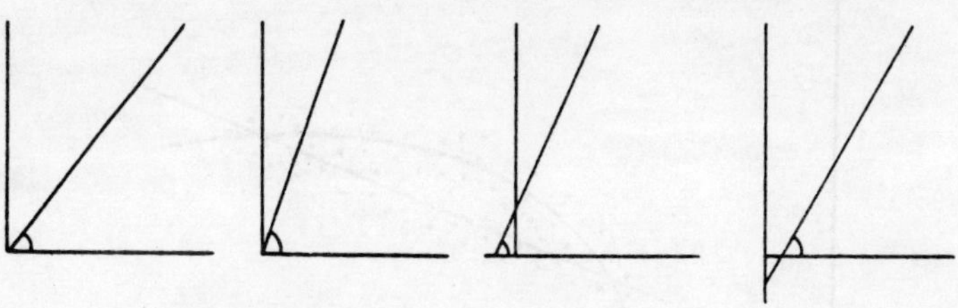

As shown on page 165, suppose mn is the line of regression. If any known value of x is taken, 'a' can also be found out by

$a = \bar{y} - b\bar{x}$ where b is the base of regression and a is the coefficient of regression.

Plot No.	Fertilizers	Yield
1	f_1	y_1
2	f_2	y_2
3	f_3	y_3
4	f_4	y_4

For any value of f which lies between f_2 and f_3 say $f_{2\cdot3}$, what would be the yield expected? It may be 2.3 or 2.5 or anything between 2 to 3.

If general trend is known e.g.

f_1 f_2 f_3 f_4 and y_1 y_2 y_3 y_4

but if actual values are to be known, it is predicted by regression which would be the most expected value.

Regression line may be a straight or a curved line e.g. a curvilinear regression curve which gives a better representation as shown below:

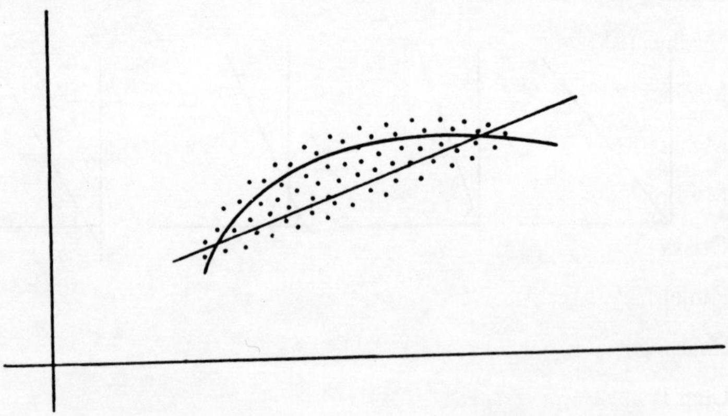

The predictions are obtained by the rules of least squares and also with the help of either correlation coefficient or the mean of the variates, variance of the variates or the individual value of the variates. The main aim is to find out as to what will a unit change in one variate will bring about a corresponding change in the other variate. This is found out with the help of regression which is depicted by $\tilde{y}$ (y tilde).

Illustration:

States	Total Population (Millions)	Population Engaged in Household Industries (Millions)			
	x	y	x^2	y^2	xy
Andhra Pradesh	35.98	1.81	1294.5604	3.2761	65.1238
Assam	11.87	0.28	140.8969	0.0784	3.3236
Bihar	46.45	1.06	2157.6025	1.1236	49.2370
Gujarat	20.63	0.55	425.5969	0.3025	11.3465
Jammu & Kashmir	3.56	0.09	12.6736	0.0081	0.3204
Kerala	16.90	0.49	285.6100	0.2401	8.2810
Madhya Pradesh	32.37	0.84	1047.8169	0.7056	27.1908
Madras	33.68	1.21	1134.3424	1.4641	40.7528
Maharashtra	39.55	0.83	1564.2025	0.6889	32.8265
Mysore	23.58	0.71	556.0164	0.5041	16.7418
Orissa	17.54	0.53	307.6516	0.2809	9.2926
Panjab	20.30	0.54	412.0900	0.2916	10.9620
Rajasthan	20.15	0.60	406.0225	0.3600	12.0900
Uttar Pradesh	73.74	1.80	5437.5876	3.2400	132.7320

West Bengal	34.92	0.49	1219.4064	0.2401	17.1108
Union Territories	7.64	0.19	58.3696	0.0361	1.4516
Total	$\Sigma x = 438.86, \Sigma y = 12.02, \Sigma x^2 = 16456.2462, \Sigma y^2 = 12.8402, \Sigma xy = 438.7868$				

$\bar{x} = 27.42875$

$\bar{y} = 0.75125$

y = a + bx in which only x is known and b can be found out with the help of the following formula:

$$b = \frac{n \sum xy - \sum x \sum y}{n \sum x^2 - (\sum x)^2}$$ in which b can be known and therefore it

can also be found out with the help of the following formula:

$$a = \bar{y} - \bar{x}\, b$$

Let us therefore first of all find out the value of b and then of a.

$$b = \frac{n \sum x\, y - \sum x \sum y}{n \sum x^2 - (\sum x)^2}$$

$$= \frac{16 \times 438.7868 - 438.86 \times 12.02}{16 \times 16456.2462 - (438.86)^2}$$

$$= \frac{7020.5888 - 5275.0972}{263299.9392 - 192598.0996}$$

$$= \frac{1745.4916}{70701.8396}$$

$$= .024,688,064$$

Now $a = \bar{y} - \bar{x}\, b$

$$= .75125 - 27.42875 \times .024688064$$

$$= .75125 - .677162735$$

$$= .074087265$$

To calculate $\tilde{y}$ values for each state, the bx, $\tilde{y}$ and r_e values have been tabulated as follows:

States	bx	$\tilde{y} = a + bx$ (Predicted Value)	$r_e = y - \tilde{y}$ (Residual Error)
Andhra Pradesh	0.88827654	0.9623638	+ 0.8476362
Assam	0.29304732	0.3671346	– 0.0871346
Bihar	1.14676057	1.2208478	– 0.1608478
Gujarat	0.50931476	0.5834020	– 0.0334020
Jammu & Kashmir	0.8788950	0.1619768	– 0.0719768
Kerala	0.41722828	0.4913155	– 0.0013155
Madhya Pradesh	0.79915263	0.8732399	– 0.0332399
Madras	0.83149399	0.9055812	+ 0.3044188
Maharashtra	0.97641293	1.0505002	– 0.2205002
Mysore	0.58214455	0.6562318	+ 0.0537682
Orissa	0.43302864	0.5071159	+ 0.0228841
Panjab	0.50116770	0.5752550	– 0.0352550
Rajasthan	0.49746449	0.5715417	+ 0.0284583
Uttar Pradesh	1.82049784	1.8945851	– 0.0945851
West Bengal	0.86210719	0.9361914	– 0.4461914
Union Territories	0.18861681	0.2627091	– 0.0727091
	Total	$\Sigma \tilde{y} = 12.0199918$	+ 1.2571656 – 1.2571574
		Total	– 0.0000082

With the help of the following checks, the lengthy calculations done above can be verified:

1) r_e i.e. $y - \tilde{y}$ should always be equal to zero. But in the above illustration the difference comes as – 0.0000082 which is almost equal to zero. This difference has occured because at some places the figures have been rounded and actual figure could not be taken.

2) $\Sigma\tilde{y}$ i.e. a + bx should come as equal to Σy i.e. in our case what should be the most ideal population engaged in household industries. In the illustration $\Sigma\tilde{y}$ and Σy are almost equal. No doubt if pairs are taken i.e. y of 1 and $\tilde{y}$ of 1, the things would differ but Σy and $\Sigma\tilde{y}$ will be equal.

To give a visual representation, a line of best fit i.e. regression line has been plotted — the values of x along the x axis and the values of $\tilde{y}$ along the y axis for each of the state in the above exercise. On plotting, the values would generate a straight line. The values of y when plotted on the axis of x and $\tilde{y}$ on the axis of y on the same graph, it would present a visual representation of the field of scatter. The distance of the points joined by drawing the lines upto the regression line and parallel to the y axis will depict the residual error r_e (Fig.22). The points falling to the left of the regression line show a positive residual error and those falling on the right hand side show a negative residual error which is statistically expressed as

$$\Sigma y - \Sigma \tilde{y} \rightarrow \text{zero}$$

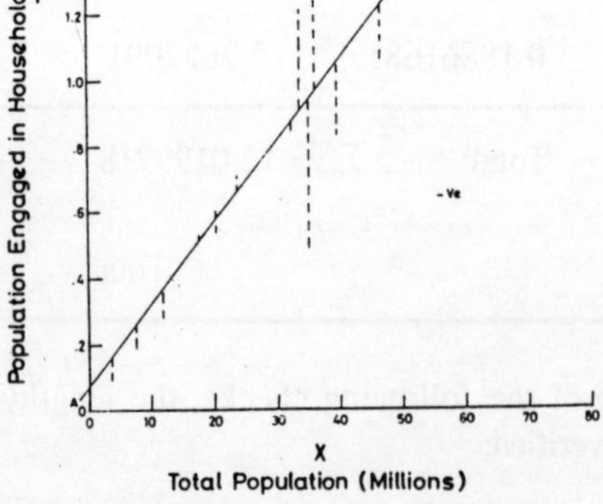

AB–Regression Line ------ Residual Error

Fig.22

CHAPTER IX

ELEMENTARY INTRODUCTION TO GEOGRAPHIC MATRICES

Matrix is a type of a box, comprising rows and columns with an array of numbers, which possesses some peculiar algebraic properties and presents a very useful technique for multivariate analysis in geography. In algebra the matrices have their largest application in the theory of simultaneous equations and linear transformations.

Concept of Preliminaries of Geographic Matrices

Geography is defined as an areal differentiation, spatial association, inter-relationship or such other factors. Each of the 'Heads' are functions of places and some aspects of characteristics and places. Geography is a function of characteristics and places.

All observations relate to a place or a location and to the characteristics of the items of study. Such observation is called geographic factor which has two components:

1. Spatial Component and
2. Attribute Component.

Spatial component relates to the locational aspect of the geographic factor. This aspect may be of the form of a single point or place or region or province e.g. the rainfall data of Delhi alone or for Agra also i.e. it may be for a series of places.

Attribute component refers to the characteristics of the item of study where the observation is made and the geographic factors recorded e.g. for any place, the rainfall, temperature, pressure etc. are recorded or they may be for a number of places. The style of record of geographical factors develops a series called geographic series.

If a single factor is taken and a single attribute of that place is considered, it makes a unit of geographic factor. There are two possibilities:

1. Geographic factor of a single attribute of a variety of places may be collected e.g. Rainfall (R) for places P_1, P_2, P_3......P_n.

This shows areal distribution which will form geographic series for a place of recording.

2. Variety of attributes of a single place may be collected e.g. rainfall of Delhi where rainfall is one factor and Delhi is one place.

There are again two possibilities. The first possibility is that if the rainfall of Delhi and various other places is recorded, it would indicate the distribution of rainfall in space or Spatial Distribution.

The other possibility is that if for Delhi, the rainfall, temperature, humidity, sunshine etc. is recorded, it shows a number of attributes for a single place. It shows a geographic series of locational inventories of a single place of a number of characteristics i.e. it provides material for geographic integeration or that it integrates various elements i.e. Locational Geographic Integration.

It shows that all geographic factors are resolved into two components — Place and Attribute.

After the collection of data pertaining to place and attribute, a tremendous collection is obtained which can be arranged in a particular form in an array or box. The above two geographic series can be combined by the matrix system.

Let there be n places and at each place m attributes are recorded. The data thus obtained may be systematically arranged in a matrix form as shown on page 173.

Suppose the data is for the whole of the world and every item is entered in the column of characteristics, then the resulting chart will be called as matrix. This box condenses all in one. This technique is called Matrix Technique and which is of great help in geography in making measurements and quantified analysis of geographical situations.

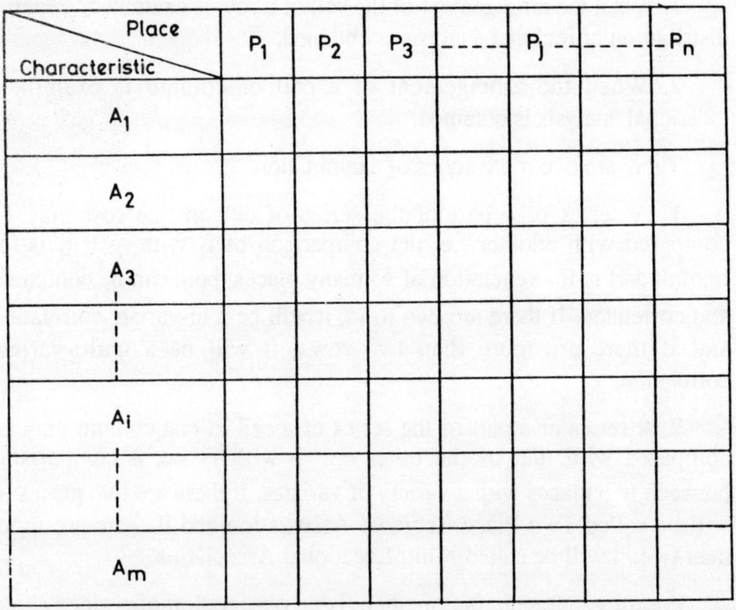

Cell is an intersection of a column and a row. ij cell is a unit geographic factor of jth place and ith characteristic. Every cell is a geographic factor. See figure below:

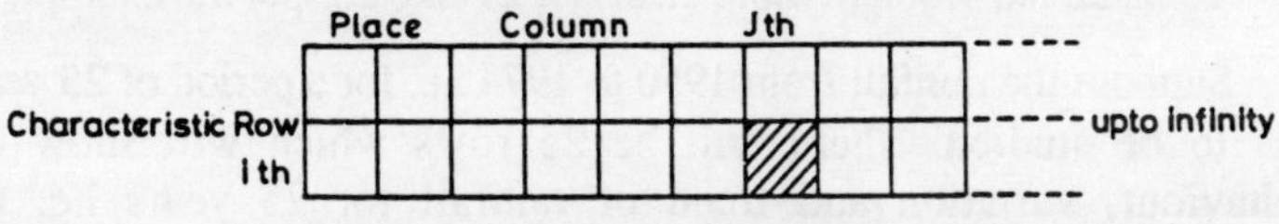

If a series of rows are taken, it means that an areal distribution is being attempted in which the place is not changing. It would give the personality of the place j in various attributes e.g. rainfall, soil, vegetation, temperature etc. which when combined give geographic personality i.e. locational studies. Rows lead to a study of region or space variation or association.

1. When the arrangement of the cell of a row is examined, a spatial distribution or regional analysis is obtained.

2. When the arrangement of a cell of column is examined, locational analysis is obtained.

There are two more styles of examination.

1. A series or a part of the series of cell in one row may be compared with another i.e. the comparison of i_1 with i_2. If i_1 is for rainfall and i_2 for vegetation of so many places, both can be compared and correlated. If there are two rows, it will be a bi-variate correlation and if there are more than two rows, it will be a multi-variate correlation.

2. A series or a part of the series of a cell in one column may be compared with that of the other e.g. j_1 with j_2 i.e. a comparison between two places with a variety of variates. If there are two places, it will be called Two Place Regional Association and if there are more than two, it will be called Multi Locational Association.

For all geographic factors there is an observation of a place about its attribute. In the matrix box composed earlier, the dimension of time was not taken into consideration. The event may happen today, tomorrow or at some other time. When the time is also taken into consideration, the analysis is called Temporal Analysis in date or time.

To make the problem more clear, let us take an specific example.

Suppose the rainfall from 1950 to 1974 i.e. for a period of 25 years has to be studied. There will be 25 rows which will show the behaviour, variation and trend of rainfall for 25 years i.e. the distribution for a place and for the number of point of time (25 years in this case) i.e. two studies in one series.

Now taking for example the style 2 possibility, we can compare a pair of cells with a single column i.e. rainfall with temperature i.e. we try to find one relationship between the rainfall and temperature. It is for localized study of two characteristics i.e. we try to find association between two geographic factors i.e. spatial association. The study will be highly localized. It may indicate that a specific rainfall and temperature is ideal for the production of a particular crop i.e. differentials have been integrated together. It would give a wider dimension of spatial association. It may be developed for various points of time say for 50, 55, 60 years etc. It would give trend, variation and behaviour for the same place, may be for a single point of time or for many points of time i.e. again two studies in one series i.e. in both styles 1 and 2, one can have four series e.g.

1. (a) Study for one particular year only,

   (b) Study for more than one year.

2. (a) Study for one particular year only and

   (b) Study for more than one year.

On the basis of the above study, therefore, a cross section in time scale has also to be made for multi period temporal analysis and entered into the form of matrix as shown below.

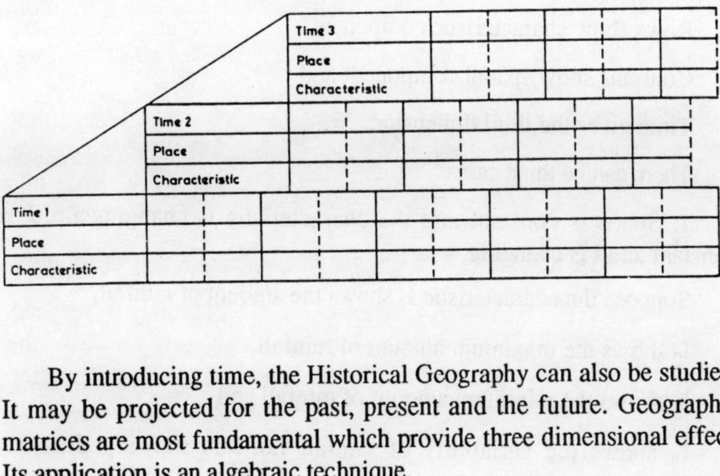

By introducing time, the Historical Geography can also be studied. It may be projected for the past, present and the future. Geographic matrices are most fundamental which provide three dimensional effect. Its application is an algebraic technique.

Further Explanation of the Entry of Three Components in Matrix

In the matrix given on page 176.

T_1, T_2, T_3............stand for different years,

J_1, J_2, J_3............stand for different places and

I_1, I_2, I_3............stand for different characteristics

and so on.

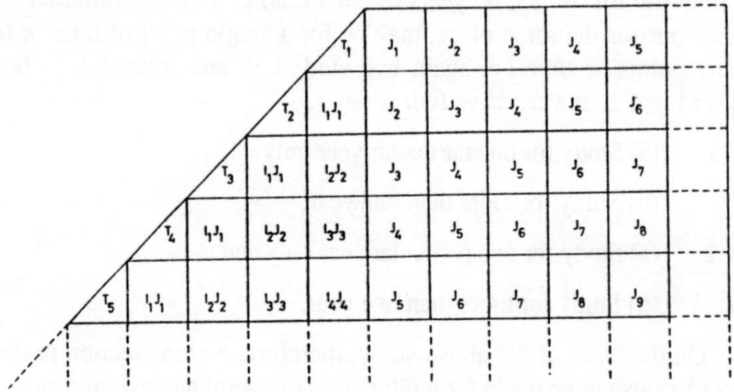

It is in a way superimposition of data of different years over one another.

Rows show characteristic component,

Columns show spatial component and

Time gives the third dimension.

There can be three cases:

1. Space is constant and the characteristic is changing i.e. J is constant and I is changing.

Suppose the characteristic I_1 shows the amount of rainfall,

I_2 shows the maximum amount of rainfall,

I_3 shows the minimum amount of rainfall and

I_4 shows the variability of rainfall i.e. I is showing different attributes of rainfall.

2. The characteristic is constant and the space is changing i.e. I is constant and J is changing.

J_1, J_2, J_3, J_4 ....J_n

I_1 A_1, A_2, A_3, A_4 ....A_n (A is attribute of characteristic).

3. The space and the characteristic both are changing and none is constant i.e. it is a combination of 1 and 2.

	J_1	J_2	J_3	J_4....................	
	A_1	A_2	A_3	A_4....................	
I_1	MA_1	MA_2	MA_3	MA_4..........	(M is amount of rainfall)
I_2	RA_1	RA_2	RA_3	RA_4............	(R is maximum rainfall)
I_3	SA_1	SA_2	SA_3	SA_4............	(S is minimum rainfall)
I_4	VA_1	VA_2	VA_3	VA_4............	(V is variability of rainfall)

But without the third dimension of time besides place and characteristic, the study of geography is incomplete because with years things are changing and also forecast and prediction is not possible. Therefore to assess and understand the trend and nature of variation in the items of study, nature of occurrence in the past is necessary. The matrix operations are very much used in the determination of correlation and principle component analysis.

The matrix can also be formed into sub-matrix for micro study which provides a more detailed study than that given by the original or primary matrix.

There may be two cases:

a) The arrangement of a pair of cells or the whole series of cells in a row.

b) The arrangement of a pair of cells or the whole series of cells in a column.

A comparison between a) and b) can be done whereas to see the changing pattern a comparison between a) and b) with respect to column can also be done.

Taking an example from population one can have a matrix as follows:

Suppose the square MNOP is for population then various aspects of population can be studied.

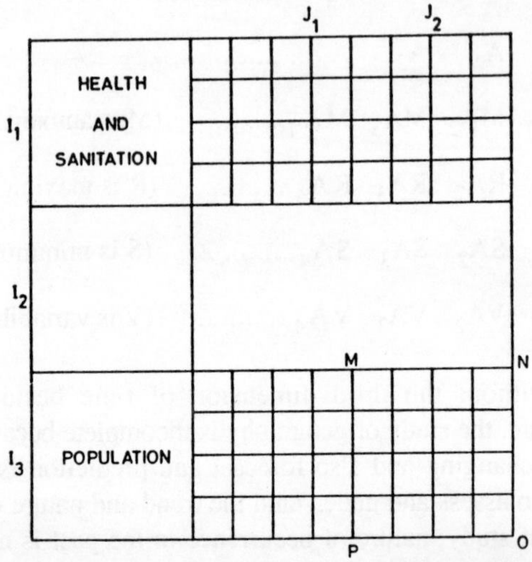

To make the problem more clear, let us enlarge the square MNOP.

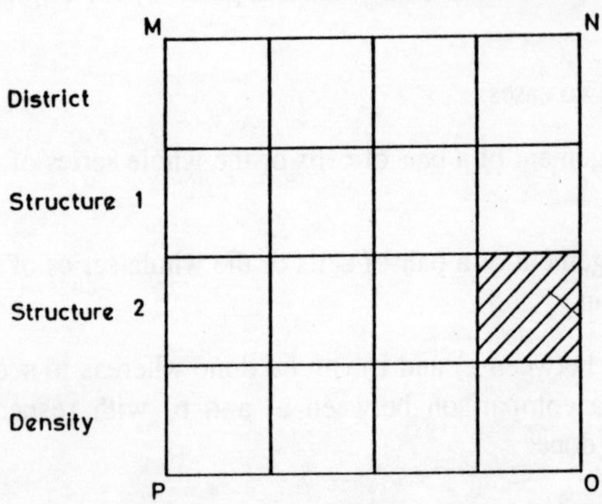

It is just possible that the sub-matrix is not contiguous as shown in A above. Here we can take pairs of J_1 and J_2 in I_1 and compare them with J_1 and J_2 in I_2 and thus find if there is any relationship showing areal distribution of geographic characteristics. In this study also the third dimension of time may be added.

Attributes cannot behave like cardinal numbers e.g. 1 and 2 in this case cannot be added to or subtracted from one another. Geography is a two dimensional matrix and it becomes three dimensional when year or time is also taken into account.

Basic Fundamentals of Matrix Algebra

Matrix is an array of number of elements arranged in a box in columns and rows and in algebra it is expressed as $[^{a}{}_{ij}]$ where a_{ij} is the (i,j)th cell entry. The order of matrix is defined by its number of rows and columns. A matrix having m rows and n columns is called a matrix of order m by n or m × n.

Each cell entry is called an element of the matrix. Matrices are of a number of types.

1. Square Matrix. A square matrix has equal number of columns and rows i.e. $[a_{ij}]$ is a square matrix if m is equal to n. If there are three columns and three rows, it means that there are 9 elements. A square matrix may be expressed as follows:

$$\begin{vmatrix} a_{11} & a_{12} & a_{13} \\ a_{21} & a_{22} & a_{23} \\ a_{31} & a_{32} & a_{33} \end{vmatrix}$$

2. Diagonal Matrix. In a square matrix if all the elements except the principle diagonal elements are zeros, it is called a diagonal matrix such as

$$\begin{vmatrix} 2 & 0 & 0 \\ 0 & 5 & 0 \\ 0 & 0 & 8 \end{vmatrix}$$

The above matrix is a squared diagonal matrix of third order. If both the diagonals have some element and the rest of the elements are zero, then it will cease to be a diagonal matrix as shown on page 180.

$$\begin{vmatrix} 2 & 0 & 3 \\ 0 & 5 & 0 \\ 8 & 0 & 2 \end{vmatrix}$$

3. Identity Matrix. This matrix is a special form of diagonal matrix in which non-zero elements are equal to one as shown below

$$\begin{vmatrix} 1 & 0 & 0 \\ 0 & 1 & 0 \\ 0 & 0 & 1 \end{vmatrix}$$

If there is any other number except 1, then it would not be an identity matrix. This identity matrix in algebra is supposed to be a unit.

4. Null Matrix. In this type of matrix all the elements are zero and it stands for what a zero stands in arithmetics e.g.

$$\begin{vmatrix} 0 & 0 & 0 & 0 \\ 0 & 0 & 0 & 0 \\ 0 & 0 & 0 & 0 \end{vmatrix}$$

It is a matrix of the order 3×4.

5. Transposed Matrix. Such a matrix is obtained by interchanging columns with rows or rows with columns. A transposed matrix is indicated as a^T or

Original Matrix — Transposed Matrix

$$\underset{3\times 4}{A} = \begin{vmatrix} 2 & 1 & 5 & 7 \\ 3 & 4 & 4 & 8 \\ 6 & 5 & 2 & 9 \end{vmatrix} \qquad \underset{4\times 3}{A^T} = \begin{vmatrix} 2 & 3 & 6 \\ 1 & 4 & 5 \\ 5 & 4 & 2 \\ 7 & 8 & 9 \end{vmatrix}$$

6. Triangular Matrix. This matrix is a special form of square matrix in which all the elements either above or below the principle diagonal are zero e.g.

$$\begin{vmatrix} 2 & 3 & 8 \\ 0 & 5 & 7 \\ 0 & 0 & 9 \end{vmatrix} \qquad \begin{vmatrix} 2 & 0 & 0 \\ 3 & 5 & 0 \\ 8 & 9 & 7 \end{vmatrix}$$

or

$$\begin{vmatrix} 0 & 0 & 2 \\ 0 & 5 & 0 \\ 8 & 0 & 0 \end{vmatrix}$$

In the above matrix, all the elements above and below the diagonals are zero. If both the diagonals are taken into consideration and all the elements apart from both the diagonals are zero, it is not a triangular matrix e.g.

$$\begin{vmatrix} 5 & 0 & 2 \\ 0 & 3 & 0 \\ 4 & 0 & 8 \end{vmatrix}$$

7. Symmetric Matrix. Symmetric Matrix is a square matrix which remains unchanged when transposed as shown below

$$\begin{vmatrix} 2 & 5 & 7 \\ 5 & 9 & 3 \\ 7 & 3 & 4 \end{vmatrix}$$

There are four basic operations in algebra and arithmetic. They are addition, subtraction, multiplication and division. Let us see as to how can they be adjusted in matrix. The simplest of them all is addition and it is well defined in matrix of the same order but it is not defined in matrix of different order. The sum of the two matrices is obtained by adding the corresponding elements provided the two matrices are of the same order as shown below

$$\begin{vmatrix} 1 & 2 & 3 \\ 4 & 5 & 6 \\ 7 & 8 & 9 \end{vmatrix} + \begin{vmatrix} 1 & 3 & 5 \\ 8 & 9 & 10 \\ 4 & 6 & 5 \end{vmatrix} = \begin{vmatrix} 2 & 5 & 8 \\ 12 & 14 & 16 \\ 11 & 14 & 14 \end{vmatrix}$$

But in A B

$$\begin{vmatrix} 1 & 2 & 3 \\ 4 & 5 & 6 \\ 7 & 8 & 9 \end{vmatrix} + \begin{vmatrix} 1 & 2 \\ 3 & 4 \\ 5 & 6 \end{vmatrix}$$

addition of A and B matrices is not possible in the above case because [A] is of the order 3×3 whereas [B] is of the order 3×2.

General or Standard form of matrix of nth order is

$$\begin{vmatrix} a_{11} & a_{12} & a_{13} & a_{14} & ....a_{1n} \\ a_{21} & a_{22} & a_{23} & a_{24} & ....a_{2n} \\ a_{31} & a_{32} & a_{33} & a_{34} & ....a_{3n} \\ a_{41} & a_{42} & a_{43} & a_{44} & ....a_{4n} \\ \cdots & \cdots & \cdots & \cdots & \cdots \\ \cdots & \cdots & \cdots & \cdots & \cdots \\ \cdots & \cdots & \cdots & \cdots & \cdots \\ \cdots & \cdots & \cdots & \cdots & \cdots \\ a_{n1} & a_{n2} & a_{n3} & a_{n4} & ....a_{nn} \end{vmatrix}$$

In the above matrix, a_{11} does not mean a eleven but it stands for a one for a_{ij}

a_{ij} means the value of a in row i and column j.

a_{11} means the value of a in row one and column one.

a_{12} means the value of a in row 1 and column two and so on.

In a way, 11 and 21 are coordinate of a point with one unit on x axis and one on y axis i.e. 21 means 2 units on x axis and 1 unit on y axis.

The multiplication of matrices is rather difficult. It is the result of 'row' times 'column' which are paired and the sum taken. Product [A] × [B] is defined only if the number of columns of [A] is equal to the umber of rows of [B]. It is also to be noted that if [A] × [B] is possible, then [B] × [A] is not necessarily possible. Also [A] × [B] is not always equal to [B] × [A].

$$\overset{[A]}{\begin{vmatrix} a_{11} & a_{12} \\ a_{21} & a_{22} \end{vmatrix}} \times \overset{[B]}{\begin{vmatrix} b_{11} & b_{12} \\ b_{21} & b_{22} \end{vmatrix}}$$

The multiplication of the above will be as follows:

Step 1. Multiply first element of row 1 in [A] with first element of column 1 in [B] we get

$$a_{11}\ b_{11}$$

Step 2. Multiply 2nd element of row 1 in [A] with 2nd element of column 1 in [B] we get

$$a_{12}\ b_{21}$$

Step 3. Add the products obtained in steps 1 and 2. This will give 1,1 element of the products and we get

$$a_{11}\ b_{11} \quad a_{12}\ b_{21}$$

Step 4. Multiply 1st element of row 1 in [A] with 1st element of column 2 in [B] we get

$$a_{11}\ b_{12}$$

Step 5. Multiply 2nd element of row 1 in [A] with 2nd element of column 2 in [B] we get

$$a_{12}\ b_{22}$$

Step 6. Add the products obtained in steps 4 and 5. This will give the (1,2) element of the product and we get

$$a_{11}\ b_{12} \quad a_{12}\ b_{22}$$

Step 7. Multiply 1st element of row 2 in [A] with 1st element of column 1 in [B] we get

$$a_{21}\ b_{11}$$

Step 8. Multiply 2nd element of row 2 in [A] with 2nd element of column 1 in [B] we get

$$a_{22}\ b_{21}$$

Step 9. Add the products obtained in steps 7 and 8. This will give the (2,1) element of the product and we get

$$a_{21}\ b_{11} \quad a_{22}\ b_{21}$$

Step 10. Multiply 1st element of row 2 in [A] with 1st element of column 2 in [B] we get

$$a_{21}\ b_{12}$$

Step 11. Multiply 2nd element of row 2 in [A] with 2nd element of column 2 in [B] we get

$$a_{22}\, b_{22}$$

Step 12. Add the products obtained in steps 10 and 11. This will give (2,2) element of the product and we get

$$a_{21}\, b_{12} \quad a_{22}\, b_{22}$$

The products are then added to get the answer which will be as follows:

$$\begin{vmatrix} a_{11}\, b_{11} & a_{12}\, b_{21} & a_{11}\, b_{12} & a_{12}\, b_{22} \\ a_{21}\, b_{11} & a_{22}\, b_{21} & a_{21}\, b_{12} & a_{22}\, b_{22} \end{vmatrix}$$

It is essential to know as to which matrix is being multiplied and which is multiplying because [A] × [B] is not the same as [B] × [A].

Multiplication of matrices is written in matrix form i.e. the answer of

$$\begin{vmatrix} a_{11} & a_{12} & a_{13} \\ a_{21} & a_{22} & a_{23} \\ a_{31} & a_{32} & a_{33} \end{vmatrix} \times \begin{vmatrix} b_{11} & b_{12} & b_{13} \\ b_{21} & b_{22} & b_{23} \\ b_{31} & b_{32} & b_{33} \end{vmatrix}$$

will be written in the form of

$$\begin{vmatrix} c_{11} & c_{12} & c_{13} \\ c_{21} & c_{22} & c_{23} \\ c_{31} & c_{32} & c_{33} \end{vmatrix}$$

The standard form is $c_{ij} = \Sigma\, a_{ik}\, b_{kj}$ where k indicates the number of rows and columns of matrix A and B.

To make the multiplication operation more clear let us take some numerical example.

$$\text{Let } A \times B \quad \begin{vmatrix} 3 & 4 & 6 \\ 2 & 1 & 5 \\ 1 & 3 & 4 \end{vmatrix} \times \begin{vmatrix} 1 & 3 & 2 \\ 2 & 3 & 1 \\ 3 & 1 & 3 \end{vmatrix}$$

It will be equal to

$$\begin{vmatrix} (3\times 1+4\times 2+6\times 3) & (3\times 3+4\times 3+6\times 1) & (3\times 2+4\times 1+6\times 3) \\ (2\times 1+1\times 2+5\times 3) & (2\times 3+1\times 3+5\times 1) & (2\times 2+1\times 1+5\times 3) \\ (1\times 1+3\times 2+4\times 3) & (1\times 3+3\times 3+4\times 1) & (1\times 2+3\times 1+4\times 3) \end{vmatrix}$$

$$\text{Or} \begin{vmatrix} 3+8+18, & 9+12+6, & 6+4+18, \\ 2+2+15, & 6+3+5, & 4+1+15, \\ 1+6+12, & 3+9+4, & 2+3+12, \end{vmatrix}$$

$$\text{Or} \begin{vmatrix} 29 & 27 & 28 \\ 19 & 14 & 20 \\ 19 & 16 & 17 \end{vmatrix}$$

SELECTED BIBLIOGRAPHY

Berry, B.J.L., 'Approaches to Regional Analysis–A Synthesis'. *Annals of the Association of American Geographers*, Volume 54, 1964. Page 2.

Birch, J.W., 'A Note on the Sample–farm Survey and its Use as a Basis of Generalised Mapping', *Economic Geography*, Volume 36, 1960. Page 254.

Blommers, P. and Lindquist, E.F., *Elementary Statistical Methods*, Oxford Book Company, 1960.

Board, C., Chorley, R.J. and Stoddart, D.R., (eds.), *Progress in Geography, Volume 1 to 7,* Butler and Tanner Ltd., 1969.

Brian, J.L., Berry, B.J.L. and Marble, D.F., (eds.), *Spatial Analysis–(A Reader in Statistical Geography)*, Prentice–Hall, Inc., 1968.

Chorley, R.J. and Hagget, P., (eds.), *Models in Geography*, Methuen and Company Ltd., 1968.

Chorley, R.J. and Hagget, P., (eds.), *Physical and Information Models in Geography*, Methuen and Company Ltd., 1969.

Cochran, G.W., *Sampling Techniques*, Asia Publishing House, 1962.

Croxton, F.E., Cowden, D.J. and Klein, S., *Applied General Statistics*, Prentice Hall of India Private Ltd., 1975.

Delton, R., Garlick, J., Minshull, R. and Robinson, A., *Networks in Geography*, George Philip and Son Limited, 1978.

Davis, J.C. and McCullagh, M., (eds.), *Display and Analysis of Spatial Data*, John Wiley and Sons, 1975.

Dickinson, G.C., *Statistical Mapping and the Presentation of Statistics*, Edward Arnold (Publisher) Ltd., 1963.

Doi, K., 'The Industrial Structure of Japanese Prefectures', *Proceedings of IGU Regional Conference in Japan, 1957*, The Organising Committee of I.G.U. Regional Conference in Japan, The Science Committee of Japan, Tokyo. 1959. Page 310.

Duncan, D., Cuzzort, R.P. and Duncan, B., *Statistical Geography (Problems in Analysing Areal Data)*, The Free Press of Glencoe, 1961.

Ebdon, D., *Statistics in Geography (A Practical Approach)*, Basil Blackwell, 1977.

Finney, D.J., *An Introduction to Statistical Science in Agriculture*, Asia Publishing House, 1965.

Fisher, R.A. (Sir), *Statistical Methods and Scientific Inference*, Oliver and Boyd, 1959.

Fraser, D.A.S., *Statistics An Introduction*, John Wiley and Sons Inc., 1958.

Gnanadesikan, R., *Methods for Statistical Data Analysis of Multivariate Observations*, John Wiley and Sons, 1977.

Goodman, R., *Teach Yourself Statistics*, English Language Book Society and England Universities Press Ltd., 1967.

Gregory, S., *Statistical Methods and the Geographer*, Longman Group Ltd., 1978.

Gregory, S., 'Models and Quantitative Techniques in Teaching (Attitudes, Opinions and Prejudices)', *Geography*, Volume 54, 1969. Page 5.

Hagget, P., *Locational Analysis in Human Geography*, Edward Arnold (Publishers) Ltd., 1965.

Hagget, P. and Chorley, R.J., *Network Analysis in Geography*, Edward Arnold (Publishers) Ltd., 1969.

Hammond, R. and McCullagh, P.S., *Quantitative Techniques in Geography: An Introduction*. Clarendon Press, 1977.

Jacobson (Jr.), P.E., *Introduction to Statistical Measures for Social and Behavioral Sciences*, The Dryden Press, 1976.

Johnston, R.J., *Multivariate Statistical Analysis in Geography*, Longman, 1978.

King, L.J., *Statistical Analysis in Geography*, Prentice–Hall, 1969.

King, L.J., 'The Analysis of Spatial Form and Its Relation to Geographic Theory', *Annals of the Association of American Geographers*, Volume 59, 1969. Page 573.

Kish, L., *Survey Sampling*, John Wiley and Sons, Inc., 1965.

Lewis, P., *Maps and Statistics*, Methuen and Co., Ltd., 1977.

Lindley, D.V. and Miller, J.C., *Cambridge Elementary Statistical Tables 3, 5, 7, 8,* 1953.

Mahmood, A., *Statistical Methods in Geographical Studies*, Rajesh Publications, 1977.

Masuyama, M., *Graphical Methods of Statistical Inference*, Maruzen Company Ltd., 1954.

Mathai, A.M. and Rathie, P.N., *Probability and Statistics*, S.G. Wasani for Macmillan Co. of India Ltd., 1977.

McCullagh, P., *Science in Geography–4, Data Use and Interpretation*, Oxford University Press, 1975.

Miller, R.L. and Kahn, S., *Statistical Analysis in Geological Sciences*, John Wiley and Sons, 1962.

Monkhouse, F.J. and Wilkinson, H.R., *Maps and Diagrams–Their Compilation & Construction*, Methuen & Co., Ltd., 1964.

Mulholland, H. and Jones, C.R., *Fundamentals of Statistics*, English Language Book Society and Butterworth, 1975.

Murthy, M.N., *Sampling–Theory and Methods*, Statistical Publishing Society, 1967.

Ostle, B., *Statistics in Research*, Iowa State University Press, 1966.

Rafiullah, S.M., 'A New Approach to Functional Classification of Towns', *The Geographer*, Volume XII, January, 1965. Page 40.

Rao, P. and Miller, R.L., *Applied Econometrics*, Prentice–Hall of India Private Ltd., 1972.

Reynold, R.B., 'Statistical Methods in Geographical Research', *The Geographical Review*, January, 1956. Page 129.

Robinson, A.H. and Sale, R.D.,*Elements of Cartography*, John Wiley and Sons, Inc., 1969.

Sadowski, W., *Statistics for Economists*, Pergamon Press, 1967.

Silk, J., *Statistical Concepts in Geography*, George Allen and Unwin, 1979.

Simpson, G. and Kafka, F., *Basic Statistics–Text Book for the First Course*, Oxford and IBH Publishing Co., 1965.

Sukhatme, P.V. and Sukhatme, B.V., *Sampling Theory with Applications*, Asia Publishing House, 1970.

Sumner, G.N., *Mathematics for Physical Geographers*, Edward Arnold (Publishers) Ltd., 1978.

Theakstone, W.H. and Harrison, C., *The Analysis of Geographical Data*, Heinemann Educational Books, 1970.

Till, R., *Statistical Methods for the Earth Scientists–An Introduction*, Macmillan Press Ltd., 1974.

Unwin, D., *Introductory Spatial Analysis*, Methuen and Co., 1981.

Walker, H.M. and Lev, J., *Elementary Statistical Methods*, Oxford and IBH Publishing Co., 1965.

Weaver, J.C., 'Crop Combination Regions in the Middlewest', *Geographical Review*, Volume 44, 1954. Page 175.

Wilks, S.S., *Elementary Statistical Analysis,* Oxford and IBH Publishing Co., 1958.

Wilson, A.G. and Kirkby, M.J., *Mathematics for Geographers and Planners,* Clarendon Press, 1975.

Wood, W.F., 'Use of Stratified Random Samples in a Land Use Study', *Annals of the Association of American Geographers,* Volume 45, 1955. Page 350.

Yeates, M.H., *An Introduction to Quantitative Analysis in Economic Geography,* McGraw–Hill Book Co., 1968.

INDEX